WITHDRAWN

MODELLING IN THE TECHNOLOGY
OF WASTEWATER TREATMENT

Related Pergamon Titles of Interest

Books:

CAIRNS: Biological Monitoring in Water Pollution
HALASI-KUN: Pollution and Water Resources
MOO-YOUNG: Waste Treatment and Utilization (Theory and Practice of Waste Management)
STUCKEY & HAMZA: Management of Industrial Wastewater in Developing Nations
SUESS: Examination of Water for Pollution Control (3 Vols)
WHO/UNEP: Waste Discharge into the Marine Environment (Principles and Guidelines for the Mediterranean Action Plan)

Journals:

Aqua (The Journal of the International Water Supply Association)
Water Research
Water Science and Technology
Water Supply (The Review Journal of the International Water Supply Association)

Full details of all Pergamon publications/free specimen copy of any Pergamon journal available on request from your nearest Pergamon office.

MODELLING IN THE TECHNOLOGY OF WASTEWATER TREATMENT

IMRE HORVÁTH C.Sc.

*Post-graduate Training Institute for Engineers and Managers
Budapest, Hungary*

PERGAMON PRESS

OXFORD · NEW YORK · TORONTO · SYDNEY · PARIS · FRANKFURT

U.K.	Pergamon Press Ltd., Headington Hill Hall, Oxford OX3 0BW, England
U.S.A.	Pergamon Press Inc., Maxwell House, Fairview Park, Elmsford, New York 10523, U.S.A.
CANADA	Pergamon Press Canada Ltd., Suite 104, 150 Consumers Road, Willowdale, Ontario M2J 1P9, Canada
AUSTRALIA	Pergamon Press (Aust.) Pty. Ltd., PO Box 544, Potts Point, N.S.W. 2011, Australia
FRANCE	Pergamon Press SARL, 24 rue des Ecoles, 75240 Paris, Cedex 05, France
FEDERAL REPUBLIC OF GERMANY	Pergamon Press GmbH, Hammerweg 6, D-6242 Kronberg-Taunus, Federal Republic of Germany

Copyright © 1984 Akadémiai Kiadó, Budapest

All rights reserved. No part of this publication may be reproduced, stored in a retrieval system or transmitted in any form or by any means: electronic, electrostatic, magnetic tape, mechanical, photocopying, recording or otherwise, without permission in writing from the publishers.

First edition 1984

Library of Congress Cataloging in Publication Data

Horváth, Imre, Dr.
Modelling in the technology of wastewater treatment.
Includes bibliographical references.
1. Sewage disposal plants—Design and construction. 2. Sewage disposal plants—Models. I. Title.
TD746.5.H68   1984   628.3   82-22530
ISBN 0-08-023978-1

British Library Cataloguing in Publication Data

Horváth, Imre
Modelling in the technology of wastewater treatment.
1. Sewage—Purification—Mathematical models.
I. Title
628'.3   TD745
ISBN 0-08-023978-1

This book is the revised English version of "A szennyvíztisztítási technológia néhány méretnövelési kérdése" published by Akadémiai Kiadó, Budapest

Translated by Z. Szilvássy
Published as a co-edition by Pergamon Press Ltd., Oxford and Akadémiai Kiadó, Budapest

Printed in Hungary

Preface

This book was written with the purpose of reviewing the potential applications of scale-up methods, similarity theory approaches and dimensional analysis to problems of wastewater treatment. The experience I have gained from 20 years of research from detailed studies of the professional literature, as well as from experimental investigation, is summarized here.

Structurally the book consists of five chapters. The first is an introduction in which the problems are formulated. This is followed by a critical review of the professional literature; this incorporates my own results and compares these with the results of other workers. Chapter 3 deals with activated sludge systems. The concept of economic similarity is introduced in Chapter 4, the potential practical applications of which will have been indicated earlier in the book. Chapter 5 gives conclusions of a general nature.

Following publication of the original Hungarian version of this book by the Publishing House of the Academy of Sciences in 1978, some additional papers and articles have appeared, including some of my more recent scientific results. Of this latest material only the information most closely related to the subject has been considered in this version, necessitating only minor expansions and amendments to the original text. Unfortunately the excellent book *Solid/Liquid Separation Equipment Scale-Up* edited by D. B. Purchas came to my notice only after completion of the manuscript. Although concentrating mainly on the field of chemical engineering, and differing radically in approach and presentation, it would have offered an interesting opportunity of further comparison in merits and methodology on a number of details, particularly in the domain of primary wastewater and sludge treatment.

This book is offered in the hope that it will provide assistance in their work to the professionals engaged in this subject.

Imre Horváth

Contents

Nomenclature 1

1. Introduction 5

2. Critical review of the literature 7

2.1 The primary stage 7
2.1.1 Screens 7
2.1.2 Sand traps 8
    Tangential sand traps 8
    Aerated sand traps 9

2.1.3 Settling basins 13
    Description of the settling process in terms of dimensionless numbers . 13
    Longitudinal-flow settling basins 16
    Circular, radial-flow settling basins 23
    Vertical-flow settling tanks 27
    Plate separators 29
    Generalization of scale-up methods 31
    The effect of turbulence 40
    Comparison of scaling-up criteria for settling basins 44

2.1.4 Filters 46
    The hydraulic aspects of scaling-up 47
    Dimensionless numbers and relationships describing filtration . 47
    Experience gained from scaling-up experiments 49
    Modelling cake filtration 51

2.2 Chemical treatment 51
2.2.1 Flocculators 51
    The zeta-potential and electrophoretic mobility 51
    The velocity gradient and energy dissipation 52
    The Camp number 53
    Modified form of the Camp number 55
    The criterion of deflocculation 57
    Turbulence and scale-up 59

2.2.2	Clarifiers	60
	Hydraulic modelling	60
	Gould's approach	61
	Technological effect of the floating sludge blanket	64
	Main clarifier dimensions	65
2.2.3	Activated-carbon adsorption devices	66
	Dimensionless correlations	67
	Applications to phenolic industrial effluent	68
	Scale-up criteria	68
2.2.4	Extraction equipment	69
2.2.5	Elutriation	69
2.2.6	Disinfection units	70
	Principles of modelling	70
	Modelling based on the flow-through characteristics	71
2.3	Biological treatment	74
2.3.1	Trickling filters	74
	Simulation of trickling filters	74
	Description by dimensionless relationships	75
	The role of turbulence	77
	Scaling-up in trickling filter technology	77
2.3.2	Aeration tanks	78
	Hydraulic similarity	78
	Flow-through studies, dispersion	86
	The effect of turbulence	89
	Energy dissipation and velocity gradient	92
	Mixing	93
	Mass transfer, oxygenation	99
	Power and energy input	117
	Similarity of bubble movement	121
	The activated-sludge aeration basin as a fermenter	122
	Description by dimensionless relations	126
	Scaling-up activated sludge systems	128
	Mathematical modelling of activated-sludge systems	130
2.3.3	Rotating-disc biological equipment	131
2.4	Sludge treatment	133
2.4.1	Digesters	133
	Reynolds numbers in non-Newtonian fluids	134
	Scaling on the basis of the modified *Re* numbers	134
	Erdmenger's concept	135
	Implications in the fermentation industry	136
	Implications in anaerobic sludge treatment	136

	Scaling based on similitude considerations and empirical data	138
	Conclusions and recommendations on digester scaling	142
2.4.2	Thickeners	145
3.	Applications of similitude in activated sludge treatment	148
3.1	Hydraulic processes	148
3.2	On modelling turbulence conditions	151
3.3	Oxygen transfer phenomena	154
3.4	Interrelations between hydraulic and oxygen transfer phenomena	159
3.5	Power consumption	162
3.6	Similarity of reaction kinetics	163
3.7	Biological similarity	164
3.8	Thermal similarity	166
3.9	Technological similarity	166
3.10	The role of recirculation	167
3.11	Partial and full similarity	169
3.12	Initial and boundary conditions	169
3.13	Scale effect	171
3.14	Comments and discussion	172
4.	Economic similarity	175
4.1	Formulation of the problem	175
4.2	Review of literature	176
4.3	Similitude and scaling considerations	176
4.4	Interpretation of the constant and the exponent a	179
4.5	Practical determination of the constants involved in the cost functions	180
4.6	Possibilities of further generalization	181
4.7	Economic similarity	182
4.8	Conclusions and recommendations	184
5.	General conclusions	186
Literature		188
Subject index		199

Nomenclature

DIMENSIONAL QUANTITIES

A	surface area;
$a, b, c, \ldots, x, y, z$	exponents (in empirical expressions);
B	width of structure;
C	concentration;
C_0, C_e	concentration of pollutants in the raw wastewater and effluent from the structure;
C_s	saturation concentration of dissolved oxygen;
C_w	resistance coefficient;
d	diameter of settling and rising particles, diameter of aerator rotor;
D	diffusion coefficient;
D_{ax}	axial diffusion coefficient;
g	acceleration due to gravity;
G	velocity gradient;
h	representative depth;
H	water depth in the structure;
K_L	mass transfer coefficient;
$K_L a$	extended mass transfer coefficient;
K_m	kinetic constant of Michaelis–Menten;
l	representative length;
L	length of structure;
M	moment, torque;
n	speed;
N	power consumption;
OC	oxygenation capacity;
p	pressure;
R	hydraulic radius;
t	time;
t_c	calculated average retention time;
T_A	surface loading rate;

T_v	volume loading rate;
x	representative variable (normally);
Q, Q_w	discharge, rate of wastewater flow;
Q_{air}	rate of air injection;
q_{air}	unit rate of air injection;
v	representative flow velocity;
v_m	mean velocity of flow;
v_p	peripheral velocity;
v_f	filtration velocity;
$\bar{v}$	average velocity (averaged over time);
v^*	pulsational velocity;
V	volume;
w	settling or rising velocity;
U_*	shear velocity;
γ	specific gravity of flowing medium;
ϱ	density of flowing medium;
ϱ_1	material density of settling particle;
η	dynamic viscosity of flowing medium;
ν	kinematic viscosity of flowing medium;
σ	surface tension;
ζ	zeta-potential;
ε	dielectric constant;
τ	shear stress.

DIMENSIONLESS QUANTITIES

A	recirculation factor;
Ar	Archimedean number;
Bo	Bodenstein number;
Ca	Camp number;
Da	Damköhler number;
E, Eu	Power number, Eulerian number;
Fr	Froude number;
Ga	Galilei number;
Ha	Hazen number;
Ho	homochronous number;
I	capillary number, hydraulic gradient;
Ly	Lyashchenko number;
MK	Mosonyi–Kovács number;
Pe	Peclet number;

Po	Poiseuille number;
R	recirculation ratio;
Re	Reynolds number;
Ri	Richardson number;
Sc	Schmidt number;
Sh	Sherwood number;
St	Stanton number, Strouhal number;
y	yield constant;
We	Weber number;
λ	scale factor, similarity transformation coefficient;
λ_x	scale factor of the variable x (e.g. $\lambda_v = v'/v''$, the scale factor of representative velocities). The single prime denotes the prototype and the double prime the model quantities.

1. Introduction

Scale-up, or some other aspect of similarity theory, may be involved in problems associated with various wastewater treatment structures:

(a) In model studies on the structures.
(b) In connection with the design, investigation of plant-scale structures of different size.
(c) In efforts to generalize the relationships, which describe the processes taking place in the structures.

The problems associated with (a) arise when trying to convert the results of model studies to prototype scale. The engineering problems under (b) are essentially of the same nature since these involve the analysis of the relationships existing between structures of different size. But, the difficulties encountered in practical design work are typical enough to warrant separate treatment. The problems in (c) are again closely related to the previous ones, especially if one wishes to describe the processes in terms of mathematics and to extend these relationships in dimensionless form to systems of different size.

The broad classification given above may include a wide variety of detailed problems, such as the development of new technology, testing newly innovated equipment, application of familiar technological processes and equipment under different conditions, applications for educational, teaching purposes, etc.

When designers of wastewater treatment facilities adopt the same type of structure or equipment with different dimensions, they are often liable to fail to make proper allowance for the minor–major changes in the technological process resulting from the change in scale. Considerations of similarity theory may enter not only the model–prototype relationship, but also when changing from a smaller, e.g. pilot installation, to a larger plant-scale unit. In the latter case the scale ratio and its inverse, the scale factor λ, will change less than when scaling-up from very small model dimensions; the scale effect will also be less pronounced. This scale effect and the failure to observe, or violation of, the conversion rules may be the sources of major dimensioning errors. This problem may occur even in cases where a successful design is adapted by apparently correct engineering methods to units of smaller or larger size. The operational difficulties resulting from such problems are usually blamed on differences in operating conditions, water quality and so on, recognizing less readily the change in size among the causes. It would be equally misleading to attribute all adverse effects to the change in scale, but this should always be considered among the potential factors.

It is pleasing to note that in recent years in the professional literature a growing number of problems have been dealt with which involve scale-up considerations related to model studies on wastewater treatment. The inconsistency of admitting the need of model tests while questioning the applicability of similarity theory has been increasingly recognized. The fact that scale-up methods fail to yield results that are fully reliable and accurate in all respects must not be misinterpreted to imply that these should be discarded altogether. As can be seen in other domains of engineering, continuous efforts are being made at the gradual improvement and development of these methods.

The application of even approximate methods of scaling-up will provide great help to the engineer in arriving at a more realistic design for the prototype structure, in contrast to the questionable approach frequently adopted in wastewater engineering, according to which the model results are accepted directly as design criteria. The rules and philosophy of similarity theory may even indicate whether to perform model, pilot or full-scale studies, or some combination in order to find the most sound and economical solution to a particular problem.

2. Critical review of the literature

This chapter will give some similarity and scaling-up problems related to wastewater treatment, together with a critical review of the professional literature on the subject.

2.1 THE PRIMARY STAGE

2.1.1 Screens

The water or wastewater entering a plant is first passed through bar and finer screens to trap the larger floating impurities. The flow conditions and local head losses in bar screens are described by Kirschmer's expression, which for straight bars is written in the form [1]

$$h_v = \beta \left(\frac{d_b}{k_b}\right)^{4/3} \sin \alpha \frac{v^2}{2g} = \zeta \frac{v^2}{2g} \tag{1}$$

where h_v is the local head loss, or differential head; d_b, the width of the rack bars in a direction perpendicular to the flow; k_b, the clear space between the bars; v, the mean flow velocity in the rack flume, in the approach cross section; α, the angle at which the rack bars are inclined to the horizontal; β, a shape coefficient, the magnitude of which depends on the bar cross section; and ζ, Weissbach's local resistance coefficient.

Starting from Kirschmer's expression—Eq. (1)—the similarity criterion for head loss across bar screens can be written within the definition range of the expression [2]

$$\lambda_\beta \approx 1 \quad \lambda_{d_b} = \lambda_{k_b} = \lambda \quad \lambda_\alpha = 1$$

Equation (1) written for two systems of different size becomes

$$\frac{v'^2}{g'h_v'} = \frac{v''^2}{g''h_v''} = Fr = \text{constant}. \tag{2}$$

The flow phenomenon considered, therefore, is characterized by the Froude number, representing the ratio of inertial force to gravity.

In terms of the Froude number, Kirschmer's expression is obtained in dimensionless form

$$\beta \left(\frac{d_b}{k_b}\right)^{4/3} Fr \frac{\sin \alpha}{2} - 1 = 0. \tag{3}$$

The condition equations of scaling-up can be found from the foregoing. The scale factor of velocities is (e.g. for $\lambda_g = 1$)

$$\lambda_v = \frac{v'}{v''} = \left(\frac{h'_v}{h''_v}\right)^{1/2} = \lambda^{1/2}, \tag{4}$$

while that of discharges

$$\lambda_Q = \frac{Q'}{Q''} = \lambda^2 \lambda^{1/2} = \lambda^{5/2}. \tag{5}$$

It would be desirable to investigate the possibility of conducting model studies on the retention of floating and suspended solids, which is the fundamental function of bars and screens. To my knowledge no results of such efforts have so far been published.

2.1.2 Sand traps

Tangential sand traps

Geiger [3] published model studies on the hydraulics of sand traps in 1942. Geiger converted the results obtained from a 1 : 10 scale model to the prototype tangential (Geiger-type) sand trap on the basis of Froude's model law. He assumed the flow to be turbulent in the model and the prototype alike, specifying a Reynolds number four times as high as the critical value.

To reproduce the settling behaviour of organic and inorganic suspended solids, Geiger departed from strict geometric similarity, in that he distorted the size of the settling particles. In the model powdered lignite and sawdust were used to reproduce the suspended sand and organic matter, blending these in the ratio 1 : 2.5. In this way he succeeded in realizing with fair approximation corresponding proportions of the deposited mineral and organic matter in the model and the prototype.

Aerated sand traps

In 1961 I developed an original analytical method in connection with hydraulic model studies on aerated sand traps. The scale-up method was first applied in designing the sand trap for the municipal treatment plant of Pécs town [4]. The scale factor of air flows introduced is given by a correspondingly modified form of Froude's law [5]

$$\lambda_Q = \frac{Q'_{air}}{Q''_{air}} = \lambda^2 \frac{w + v''_m \lambda^{1/2}}{w + v''_m} = \lambda^2 \frac{w + v'_m}{w + v'_m \lambda^{-1/2}} \tag{6}$$

where Q_{air} is the rate of air flow introduced; w, the mean rising velocity of the air bubbles; and v_m, the mean velocity of water flow in the vicinity of the air nozzle.

Equation (6) was derived originally for modelling aeration tanks with submerged air nozzles (for additional details see Chapter 3), but experiments have shown it to be reasonably applicable to aerated sand traps as well. Substituting $w \approx v'_m \approx 30$ cm/s as a value typical of operating sand traps, the simplified expression

$$\lambda_{Q_{air}} = 2 \frac{\lambda^2}{1 + \lambda^{-1/2}} = \frac{2}{1 + \lambda^{1/2}} \lambda^{5/2} = C\lambda^{5/2} \tag{7a}$$

is obtained, with the coefficient C representing the impossibility of reproducing the rising velocity of the air bubbles by observing Froude's law (the validity of which would imply that $C = 1$). Relating in the model and prototype the rate of air flow to unit tank length, Eq. (7a) assumes the form

$$\lambda_{Q_{air}} = 2 \frac{\lambda}{1 + \lambda^{-1/2}} = \frac{2}{1 + \lambda^{1/2}} \lambda^{3/2} = C\lambda^{3/2}. \tag{7b}$$

The curves representing Eqs (7a, b) are given in Fig. 1, together with the curve given by Froude's law [5].

Note that it is difficult to evaluate v_m numerically in the prototype, but it is easily seen from the model. Any observational error in the magnitude of v_m results in only a moderate error in the value of $\lambda_{Q_{air}}$ so long as λ is kept less than 10, and, therefore, remains in the range normally adopted for sand trap models. This is shown in Fig. 2, representing the $\lambda_{Q_{air}}$ vs v''_m relationship for values of 2, 5 and 10 (with $w = 30$ cm/s). As can be seen from the diagram, for $\lambda = 2$ the influence of v''_m on the value of $\lambda_{Q_{air}}$ is insignificant. For instance, when using a value of $v''_m = 35$ cm/s instead of the correct value of 30 cm/s, the resulting error in $\lambda_{Q_{air}}$ is only 5%.

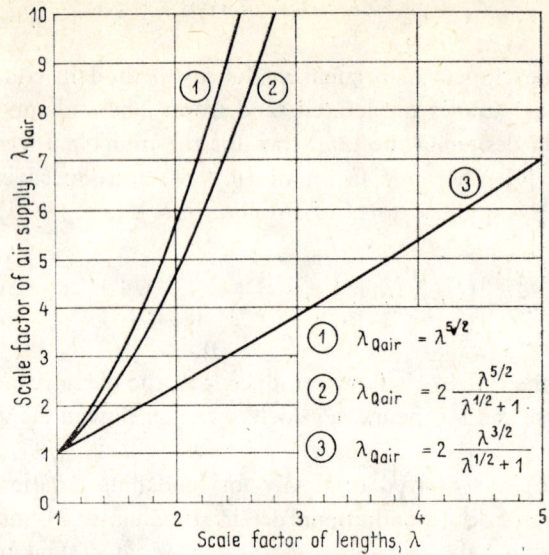

Fig. 1. Scaling air supply rates to compressed-air sand traps ($w = v'_m = 30$ cm/s)

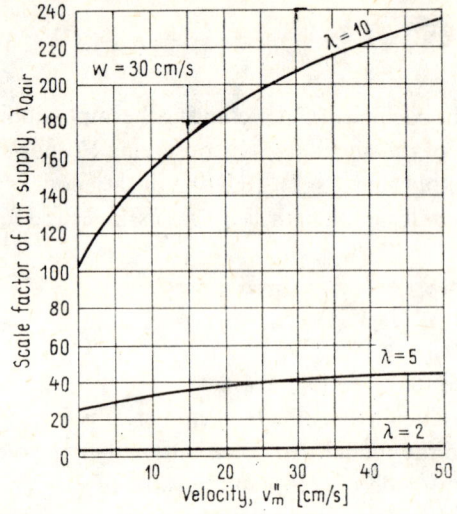

Fig. 2. $\lambda_{Q\text{air}}$ vs v''_m relationship (the parameter is λ)

A detailed discussion of the above has been presented together with the corresponding design diagrams in a report published in 1972 [6]. The diagrams suggested for dimensioning are given in Figs 3a–c. It must be emphasized that these apply to a circular tank of $D_H = 200$ cm diameter.

For other tank sizes the results must be scaled up (or down) using Fig. 1, or Eq. (7). Note that these diagrams have been derived experimentally and by considerations based on dimensional analysis.

In recent years several experimental results have been published on the scale factor of the rate of air flow discharged into aerated sand traps. As an example of the work in Hungary, the interesting series of data determined by Sallay under the guidance of Salamin [7] is given, the detailed analysis of which has been presented in more recent studies [5, 6].

Albrecht [8] reported on hydraulic studies related to a new tank design. For the alternative shown in Fig. 4, he derived, from experimental results, the expression

$$K = \frac{Q_{air} h^2}{L v^2 h_B} \tag{8}$$

where K is an experimental constant, the magnitude of which depends on the design and position of the air discharge device; Q_{air} the rate of air flow; v, the mean flow velocity under the baffle; L, the tank length; h, the depth of air diffusion; and h_B, the height of the opening under the baffle.

Dimensional analysis has led Albrecht to the following form of Eq. (8)

$$K^* = \frac{Q_{air} h^2}{L v^2 h_B} \frac{g}{v} \tag{9}$$

where K^* is a dimensionless quantity. The mean velocity characteristic of the hydraulic conditions becomes

$$v = h \sqrt{\left(\frac{Q_{air}}{K L h_B}\right)} = h \sqrt{\left(\frac{Q_{air} g}{K^* L h_B v}\right)}. \tag{10}$$

Continuing Albrecht's line of reasoning, Eq. (9) shows interesting similarity and scale-up implications. The expression can be rewritten into the form

$$\frac{1}{K^*} = \frac{vv}{gh^2} \frac{L h_B v}{Q_{air}} = \frac{vv}{gh^2} \frac{Q_w}{Q_{air}} = \frac{Fr}{Re} \frac{Q_w}{Q_{air}} \tag{11}$$

where Q_w is the water flow passing the area $L h_B$. From Eq. (11) it can be seen that this particular flow phenomenon is characterized by the ratio of the Froude and Reynolds numbers. (In the Hungarian literature on seepage theory the Fr/Re ratio is known as the Mosonyi–Kovács number.) Therefore, it follows that the forces of gravity and friction are the predominant forces. This is believed to be acceptable as a crude approximation, since the effect of the inertial force is also important (as expressed implicitly by the ratio Q_w/Q_{air}).

Assuming geometrically similar systems further in the case of

$$\lambda_v = \lambda_g = 1$$

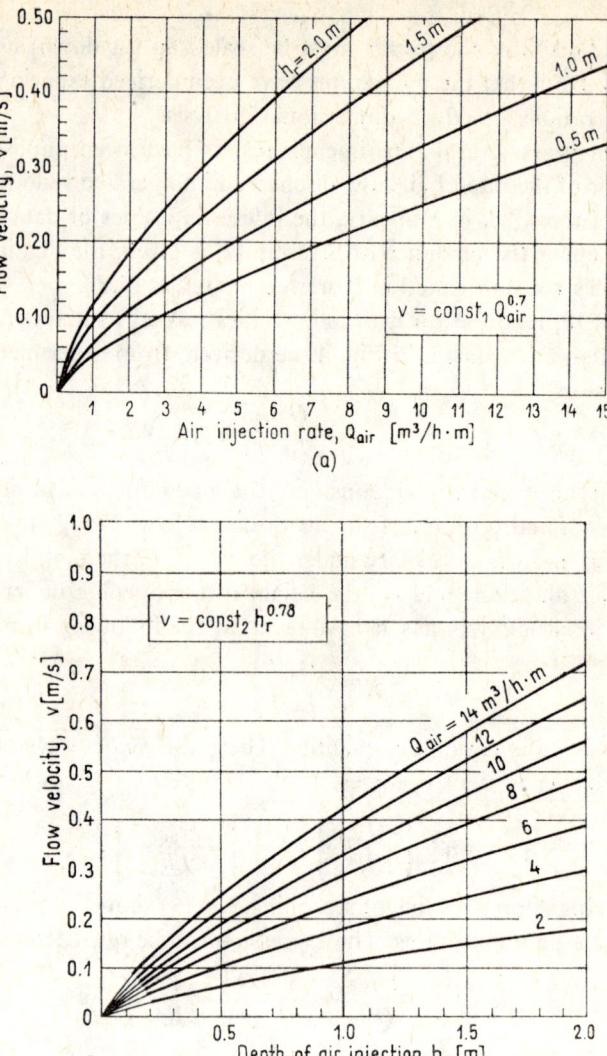

the scale factor of air flow rates becomes, with the dimensionless quantity of Eq. (11)

$$\lambda_{Q_{air}} = \lambda_v^2 = \lambda_v \lambda^{-2} \lambda_{Q_w} = \lambda_{Q_w}^2 \lambda^{-4}. \tag{12}$$

It is useful to ensure invariance simultaneous of $1/K^*$ and the Fr number, so that from Eq. (12)

$$\lambda_{Q_{air}} = \lambda. \tag{13}$$

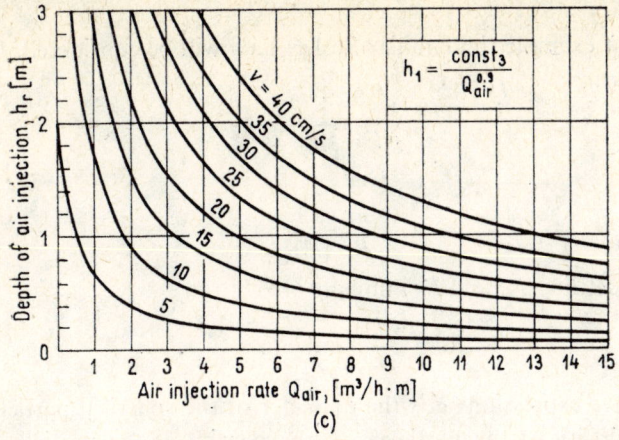

Figs 3(a)–(c). Dimensioning graphs for compressed-air sand traps (diameter of structure $D_H = 2.0$ m)

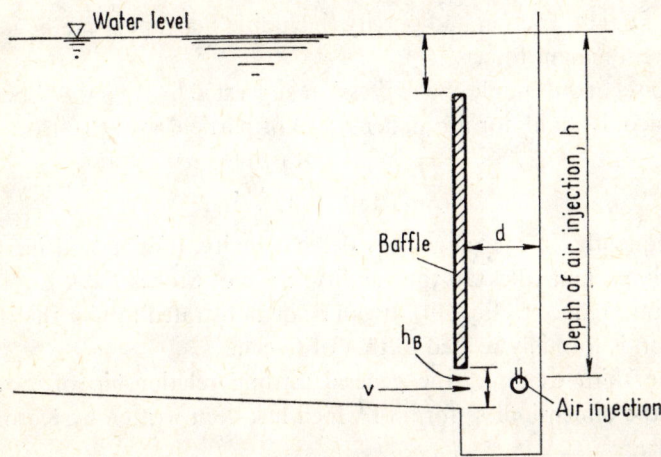

Fig. 4. Explanatory diagram for the Albrecht-type aerated sand trap

Equation (13) agrees with Eq. (7a) derived with more general validity in the case of $\lambda = 1$ alone. It can be concluded that the use of Eq. (13) is justified only in the cases where $\lambda \approx 1$, within the experimentally established validity range of Eq. (11).

2.1.3 Settling basins

Description of the settling process in terms of dimensionless numbers

Before embarking upon an examination of the scale-up problems associated with various types of settling basin, it is worthwhile to re-state briefly the description of the settling processes by dimensionless numbers.

As a first example the familiar Stokes Law will be considered. In dimensionless form

$$\frac{wv}{gd^2} = \frac{1}{18}\frac{\varrho_1-\varrho}{\varrho} \qquad (14a)$$

or

$$\frac{Fr}{Re} = \frac{1}{18}\varrho^* \qquad (14b)$$

where the settling Re and Fr numbers are

$$Re = \frac{wd}{v} \quad \text{and} \quad Fr = \frac{w^2}{gd}.$$

In the above expressions d is the diameter of the spherical particle; w, the settling velocity of the particle; ϱ_1, the density of the particle material; ϱ, the density of the medium; v, the kinematic viscosity of the medium; and g, gravitational acceleration.

From Eq. (14) friction and gravity (including buoyancy) can be identified as the predominant forces.

As the second example the expression suggested by Rubey will be considered, which is valid for a broader range of particle sizes (0.002–10 cm) [9]

$$\frac{Fr}{Re} = \frac{1}{18}\varrho^* - \frac{1}{12}Fr. \qquad (15)$$

Consequently, in the range considered, gravity, friction and inertia play equally important roles. In the validity range of Stokes Law, Eq. (15) can be shown to reduce to Eq. (14). It can be demonstrated further that Rubey's expression is formally related to that of Oseen.

As the third example, the general settling relationship of Newton is quoted, the dimensionless form of which has been written by Kármán (see [291]) as

$$C_w Re = \frac{4}{3}\frac{gd^3(\varrho_1-\varrho)\varrho}{\eta^2}. \qquad (16a)$$

By introducing additional dimensionless numbers the following alternative forms are obtained

$$C_w Re = \frac{4}{3}Ar = \frac{4}{3}Ga\varrho^* = \frac{4}{3}\frac{Re^2}{Fr}\varrho^*. \qquad (16b)$$

Lyashchenko has suggested a method compatible with Eq. (16) for calculating the settling velocity in terms of dimensionless quantities including [10] the Reynolds number

$$Re = \frac{wd}{v}, \qquad (17)$$

the Archimedes number (expressed with the help of the Galilei number)

$$Ar = Ga\varrho^* = \frac{Re^2}{Fr}\varrho^* = \frac{gd^3(\varrho_1-\varrho)\varrho}{\eta^2} \qquad (18)$$

or the Lyashchenko number

$$Ly = \frac{Re^3}{Ar} = \frac{FrRe}{\varrho^*} = \frac{w^3\varrho^2}{\eta(\varrho_1-\varrho)g}. \qquad (19)$$

Using these dimensionless numbers the full range of settling velocities can be described for the (unhindered) spherical particles case.

From the above examples it can be inferred that the settling velocity of spherical particles can be described with the help of Froude and Reynolds numbers and the density ratio ϱ^*. Consequently, the scale-up conditions are also controlled by these dimensionless numbers, with due regard to the velocity limits applying to a particular case. The three predominant forces may vary in relative importance in particular cases. Thus in the validity range of Stokes Law, the effect of the inertial force is negligible and then in the case of

$$\lambda_\nu = \lambda_{\varrho_1} = \lambda_\varrho = \lambda_g = 1$$

the condition equation of similarity is

$$\lambda_w = \lambda_d^2. \qquad (20)$$

In a more general case Fr and Re would have to be entered simultaneously in writing the conditions of scaling-up (as when applying Rubey's relationship for example). The familiar theoretical and practical difficulties involved can, however, be overcome only under special conditions.

When applying Newton's Law, the condition equation of scaling-up is in the case of

$$\lambda_\eta = \lambda_{\varrho_1} = \lambda_\varrho = \lambda_g = 1$$

obtained as

$$\lambda_w = \lambda_d^{1/2}\lambda_{C_w}^{-1/2} \qquad (21)$$

where allowance must be made for the fact that C_w depends both on Re and the shape factor (e.g. the sphericity ξ). Evidently, by introducing practical, modelling considerations, Eq. (21) can be expanded further, e.g. by specifying $\lambda_{\varrho_1} \neq 1$, or some other condition. It should be noted that the role of ϱ^* is comparable in importance to that of Fr and Re in modelling computations. Reference is made in this respect to the theoretical similarity investigations of Field [11] related to sediment transport. The conclusions arrived at by dimensional analysis are supported by experimental results stressing the need of including the density ratio ϱ^* in reproducing sediment settling processes. Bewtra has also adopted dimensional analysis as the basis in tracing the diagrams relating C_w, Re and ϱ^* to facilitate settling velocity computations [12].

Longitudinal-flow settling basins

Of the various settling facilities, longitudinal-flow, rectangular settling basins have been dealt with most extensively in the literature on applied similarity theory. The overall impression is that most investigators have adopted an approach based on the Froude and Reynolds numbers.

One of the major problems when using the Froude Law consists of measuring the velocities in the model. This difficulty is often overcome by the "flow-through", or "passage" methods, involving the injection of some tracer substance. Some authors have specified the invariance of the Froude number to hydraulic gradient ratio Fr/I as the similarity criterion, especially in cases where appreciable hydraulic losses occur. In this respect the similarity of roughnesses in the model and the prototype is an essential consideration.

Flow under pressure has occasionally been realized in order to make the model relationship of Reynolds applicable. Following a suggestion by Averkhyev (see Shifrin [13]) the water surface in the basin is replaced in this case by a plane transparent plate. Both air and water can be used as the flowing medium in such models. The scale factor λ is adopted preferably in a way to create flow conditions falling into the turbulent range termed "self-modelling". This is achieved if [14]

$$Re'' \geq Re_{s\,min} \qquad (22)$$

where $Re_{s\,min}$ is the lowest possible Reynolds number of the self-modelling range of turbulent flow.

The experimental results of Averkhyev have shown $Re_{s\,min} = 1500$–2000 to be advisable in settling basin studies. It should be remembered, however, that the value of the $Re_{s\,min}$ is materially influenced by design particulars, such as the entrance to the basin.

The results obtained by Arent [14] in a pressurized air-flow model of rectangular settling basins confirmed the principle expressed by Eq. (22).

Further experimental proof of the applicability of Eq. (22) was derived by Medvedev from the results of open-flow hydraulic model tests on rectangular settling basins. Medvedev has also pointed out the necessity of checking experimentally the possibility of self-modelling of the hydraulic phenomena under consideration. This can be accomplished, for instance, by determining the velocity distributions in representative cross-sections at different Re, the similarity of the velocity patterns demonstrating the existence of the self-modelling range.

Medvedev [15] performed actual measurements with clear water at constant temperature at $Re''=2000$ and 3000, corresponding to $Q=6.9$ and 10 l/s and $v=6.9$ and 10 mm/s, respectively. From the fair agreement between the velocity distributions obtained he inferred the existence of the self-modelling range, further that the criterion of Eq. (22) was satisfied. The experimental data suggested the advisability of adopting the range $Re''\geq 2000$–3000. Moreover, he called attention to the fact that model studies on the performance of settling basins yield results satisfactory from all respects when the actual sewage and suspended solids are used. Additional information on scaling-up problems can be derived from registering the surface profile of the deposits, the variations in removal efficiency, etc.

Starting from the consideration that in settling basins the Froude number is low ($Fr<0.005$), Levi [16] suggested its omission altogether as a similarity criterion in model tests concerned with the hydraulics of the structure. However, in systems expected to remove suspended solids, additional similarity criteria enter the picture. Based on the work of Velikhanov, Nikitin and others (see Levi [16]) specified the following scaling-up criteria to be observed in the model studies on settling basins:

(a) $$Fr_1 = \frac{v^2}{gh} = \text{constant} \tag{23a}$$

$$Fr_2 = \frac{w^2}{gh} = \text{constant} \tag{23b}$$

whence

$$\lambda_v = \lambda_w = \lambda_h^{1/2} \tag{24}$$

(b) $\quad\quad Re'' = Re_{\text{crit}} = 150 \quad\quad Re' > Re_{\text{crit}} = 150 \quad\quad$ (25a,b)

(c) $\quad\quad Re'' = Re' \quad\quad Re' < Re_{\text{crit}}.\quad\quad$ (26a,b)

In the above: $Re = vd/v$, d is the diameter of the suspended particles, v, a characteristic flow velocity, mean velocity; w, the settling (fall) velocity; and Re_{crit}, the critical Reynolds number at the lower limit of turbulent flow.

In (b), where $Re' > Re_{\text{crit}} = 150$

$$\lambda_d = \frac{1}{\sqrt[3]{\lambda_{q_t}}} \left(\frac{150}{Re'}\right)^{2/3} \tag{27a}$$

while in (c), where $Re' < Re_{crit}$

$$\lambda_d = \frac{1}{\sqrt[3]{\lambda_{\varrho_1}}}. \tag{27b}$$

In a rectangular settling basin of depth H and length L the scale factor of settling velocities ($w = H/Lv$), including also the possibility of distortion, is

$$\lambda_w = \lambda_h \lambda_l^{-1} \lambda_v. \tag{28a}$$

To allow for the effect of turbulent flow, the empirical exponent 2/3 was suggested by Velikhanov and Levi [16] for the ratio λ_h/λ_l, thus

$$\lambda_w = \lambda_v \left(\frac{\lambda_h}{\lambda_l}\right)^{2/3} \tag{28b}$$

reducing in geometrically similar (undistorted) systems to $\lambda_w = \lambda_v$.

This is a good point to refer to the results of Gordanov, who perfected the modelling techniques of the settling processes in wastewater containing suspended solids [17]. The critical analysis of Levi's method revealed the difficulties that arise in its practical application. He cited examples to demonstrate the virtual impossibility of simultaneously observing the criteria of Eqs (24) and (27a) or (27b). For instance, in the case of $\lambda_h = 4$, the factors $\lambda_v = \lambda_w = 2$, with the criterion $\lambda_d = \lambda_h$. Eq. (27b) gives $\lambda_{\varrho_1} = \lambda_{\gamma_1} = 0.0156$. Thus if $\varrho'_t = 2.65$ g/cm³, then $\varrho''_t = 170$ g/cm³, which is obviously impossible in practice. Moreover, any attempt at simultaneously observing Eqs (23a, b) and (26a, b)—the Fr and Re criteria—is bound to face obstacles of the same kind.

Gordanov derived by theoretical analysis a set of similarity criteria for rectangular, longitudinal-flow settling basins, using the general set of differential equations describing two-phase flow published by Frankl [18]. Omitting the detailed derivation, the similarity criteria he arrived at are as follows:

(a) $\dfrac{vw}{gH} = $ constant (b) $\dfrac{H}{L} = $ constant

(c) $\dfrac{w^2}{gH} = $ constant (d) $\dfrac{v^2}{gH} = $ constant (29a–f)

(e) $\dfrac{P}{\gamma_t H} = $ constant (f) $\dfrac{P}{\gamma_t L} = $ constant

where P is the hydrodynamic pressure head.

These criteria are impossible to satisfy simultaneously. By quantitative and qualitative analysis the quantities (29a, b, d) can be shown to predominate. On the basis of the invariant group (d)

$$\lambda_v = \lambda_h^{1/2}. \tag{30a}$$

Further

$$\lambda_Q = \lambda_F \lambda_v = \lambda_l \lambda_h \lambda_h^{1/2} = \lambda_l \lambda_h^{1.5}. \tag{30b}$$

In combination with Eq. (28a)

$$\lambda_w = \lambda_v \frac{\lambda_h}{\lambda_l} = \frac{\lambda_h^{1.5}}{\lambda_l} \tag{31a}$$

or with the empirically corrected Eq. (28b)

$$\lambda_w = \lambda_v \left(\frac{\lambda_h}{\lambda_l}\right)^{2/3} = \lambda_h^{1/2} \left(\frac{\lambda_h}{\lambda_l}\right)^{2/3} = \sqrt[6]{\frac{\lambda_h^7}{\lambda_l^4}}. \tag{31b}$$

According to Gordanov's approach scaling-up is accomplished essentially on the basis of Eqs (30a, b)—the Froude relationship—and Eq. (31b), satisfying approximately the criterion of hydrodynamic similarity. Note that this approach also implies that to reproduce the settling process it is sufficient to observe the similarity of the paths of the suspended particles and the settling velocity. The γ_l and particle diameter d are included implicitly in the settling velocity.

Should this approach result in inconvenient model dimensions (or other relevant conversion factors), a repeated attempt based on Eqs (22) and (28b) was suggested by Gordanov. This implies using Reynolds model law, where

$$\lambda_v = \lambda_h^{-1}. \tag{32}$$

Further from Eq. (28b)

$$\lambda_w = \lambda_v \left(\frac{\lambda_h}{\lambda_l}\right)^{2/3} = \frac{1}{\sqrt[3]{(\lambda_h \lambda_l^2)}}. \tag{33}$$

In accordance with Eq. (22) this is the case of scaling-up in the range of self-modelling.

The use of ratios $\lambda_l/\lambda_h \leq 5$ has been suggested in similar problems [19]. For practical modelling, reference is made to Zrelov [20] on the similarity criteria related to floating debris and the settling of suspended solids.

The equations describing the flow of two-phase media have been adopted by Tesaker [21]—just as by Gordanov—as the starting basis in studying the similarity criteria of flowing suspensions. Having analysed the transport and settling behaviour of the suspended matter, he succeeded in deriving a scale-up method based on the earlier results of Keulegan, Middleton and others (see Bogárdi [22]). The method proved successful in practical applications, as shown by a numerical example. Industrial effluent from an ore mine was discharged into a lake. For this case the possibility of modelling the movement of the two-phase fluid could be demonstrated positively. For the density currents studied, the model laws were found to apply, provided that the settling velocity of the suspended particles was small relative to the flow velocity. The results obtained were fully satisfactory for practical purposes, although the similarity was but a partial one. Many interesting useful pointers are to be found in the paper on distorted and undistorted models, as well as on the scale effect.

It should be noted that the literature on sediment transport contains abundant additional information and analytical methods that can be used advantageously with proper judgement in reproducing the phenomena taking place in settling basins. Reference is made first of all to the comprehensive book of Bogárdi [22]. The balance equations of sediment transport have recently been published by Bogárdi and Szűcs [23], improving by their method Frankl's idea. The results of Rouse [24], Yalin [25, 26], Komura [27], Herbertson [28], Zwamborn [29], as well as Einstein and Chien [30], besides advancing the theory of sediment transport, may also provide guidance in scaling-up treatment equipment intended to remove suspended solids.

These methods are founded without exception on a conventional approach to hydromechanics. The study to be described is based rather on the concepts and principles common to the practice of water and wastewater treatment technology and is therefore of particular interest.

Thompson [31, 32], in his doctoral thesis published in 1967, described a scale-up method for rectangular settling basins containing many novel features. By dimensional analysis, the removal efficiency of settling basins can be described in the general case with the help of the following dimensionless groups

$$\frac{C_e}{C_0} = f_1\left(\frac{C_0}{\varrho}, \frac{w}{\sigma_w}, \frac{\varrho w L}{\eta}, \frac{Q}{wLB}, \frac{w^2}{gD}, \frac{B}{L}, \frac{D}{L}, \frac{H}{L}\right) =$$

$$= f_2\left(\frac{C_0}{\varrho}, \frac{w}{\sigma_w}, \frac{\varrho Q}{\eta D}, \frac{Q}{wLB}, \frac{Q^2}{gD^5}, \frac{B}{L}, \frac{D}{L}, \frac{H}{L}\right) \quad (34)$$

where C_0, C_e the concentrations of suspended solids in the inflow to, and effluent from, the basin; L, B, D are the length, width and depth of the settling basin, respectively; H, the differential elevation between the inflow and outflow weir crests; and σ_w, the statistical parameter characterizing the distribution of the settling velocity w.

The following dimensionless groups are involved in the above expression:
Reynolds number

$$Re = \frac{wL}{\eta/\varrho} \propto \frac{\varrho Q}{\eta D} \tag{35}$$

Froude number

$$Fr = \frac{w^2}{gD} \propto \frac{Q^2}{gD^5} \tag{36}$$

and Hazen number

$$Ha = \frac{Q}{wLB} = \frac{Dv}{Lw} = \frac{T_s}{w}. \tag{37}$$

Equation (37) represents Hazen's concept founded on the surface load T_s.

To check the validity of the model relationship, Thompson studied two geometrically similar models (denoted A and B) built to different scales. The water used in the experiments contained suspended solids of 100–1500 mg/l concentration. For A and B the surface load ($T_s = Q/LB$) vs the C_e/C_0 ratio relationship is shown in Fig. 5. An important conclusion that can be arrived at from the figure is that the data obtained for geometrically similar models have similar curves. The difference between the experimental and theoretical curves is attributable in this particular case to the approximation involved in Hazen's concept. For example, the effect of turbulence can be explained on the basis of the Dobbins–Camp theory. For additional details see the subsection: The effect of turbulence, Hazen's criterion can

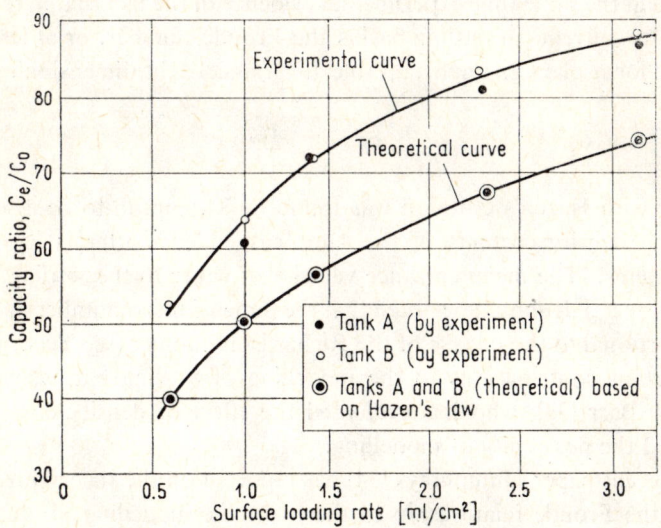

Fig. 5. The surface loading rate vs the ratio C_e/C_o for rectangular settling tanks

therefore be adopted as the condition equation of similarity, according to which

$$\lambda_Q = \lambda_w \lambda_L \lambda_B \qquad (38a)$$

or

$$\lambda_v = \lambda_w \frac{\lambda_L}{\lambda_D}. \qquad (39a)$$

Further in geometrically similar systems

$$\lambda_Q = \lambda_w \lambda^2 \qquad (38b)$$

and

$$\lambda_w = \lambda_v. \qquad (39b)$$

The relationship given by Eq. (39a) corresponds to the conventional criteria of Eq. (28a) considered earlier [the effect of turbulence can be taken into consideration also on the basis of Eq. (28b)]. Since λ_w was unity in Thompson's experiments, the condition equation of scaling-up is obtained, instead of from Hazen's criterion directly from the surface load $T_s = Q/LB$, according to Fig. 5. In this case the criteria of Froude and Reynolds' criteria can be neglected as demonstrated by the experimental results. It should be emphasized that the objective of these model experiments was the reproduction of identical C_e/C_0 ratios, and in this connection with the same removal efficiencies, rather than the realization of corresponding flow conditions or patterns. These are aspects lending novelty to Thompson's investigations compared with the scale-up methods outlined above. Thompson has provided at the same time experimental evidence of the fact that in reproducing density currents in settling basins the Froude number, or at least some modified form, plays an important role. Invariance of the dimensionless group

$$\frac{\varrho}{C_0} Fr = \frac{\varrho}{C_0} \frac{v_{in}^2}{gL} \qquad (40)$$

together with Hazen's criterion was found by Thompson to be essential in modelling density currents, if the same suspended matter is involved in both systems. (The mean entrance velocity of water to the settling basin is denoted by v_{in}). It should be noted that the dimensionless number in Eq. (40) is proportional to the inverse of the Richardson number (see Section 2.2.2).

Important contributions to the modelling of settling basins have been made by Barr [33], who has evaluated the effect of density currents and examined the possibility of modelling.

In a recent paper Humphreys [34] has reported on the successful application of the Froude relationship to the hydraulic modelling of rectangular settling basins.

Circular, radial-flow settling basins

The similarity problems associated with circular, radial-flow settling basins have been studied in detail experimentally by Clements and Khattab [35, 36]. The settling basins studied were approximately similar in geometry. The criterion of hydraulic similarity was Reynolds' model relationship in the form

$$Re = \frac{vd}{v} = \frac{Q}{\pi(R+R_{in})v} \qquad (41)$$

where $d = R - R_{in}$

$$v = \frac{Q}{\pi(R^2 - R_{in}^2)}. \qquad (41a)$$

Further R is the radius of the circular settling basin; and R_{in}, the distance of the inlet devices (e.g. Stengel nozzles) from the centre.

The experiments were performed at Re ranging from 2510 to 2700 (turbulent range) and from 1556 to 1675 (transition range), at $T = 18$ °C and using powdered aluminium oxide as the suspended matter.

The alternatives studied are shown in Fig. 6a. Figures 6b and 6c represent the settling efficiency vs time ratio relationship determined experimentally. (The time ratio has been defined by the authors as the ratio of the effective flow-through time to the effective settling time.)

The alternative designs have been numbered I–V. Figure 6b represents operating conditions, where the bottom velocities have not caused a scouring effect, while in Fig. 6c they exceeded the critical pick-up value. The correlation factors in these two alternatives were 0.95 and 0.94 respectively.

The main conclusions concerning scale-up and modelling may be summarized as follows:

(a) Model–prototype conversions were most accurate at Re near 2600.

(b) In hydraulic modelling based on the time ratio, flow conditions could be reproduced satisfactorily by using the Re number as the scale-up criterion.

(c) Similarity of settling depends more on the inflow condition than on the geometric design of the whole basin. This is reflected also by Fig. 7, showing the time ratio vs inlet ratio relationship for two models of different size. The difference between the two curves is due probably to the scale effect and implies that in the smaller model a higher time ratio and a correspondingly higher settling efficiency can be achieved over a wide range.

(d) As to be seen from Figs 6b, c, the time ratio needed for obtaining the same settling efficiency may serve as the basis of the scale-up condition equation. To reproduce the settling efficiency, the ratio of the flow-through

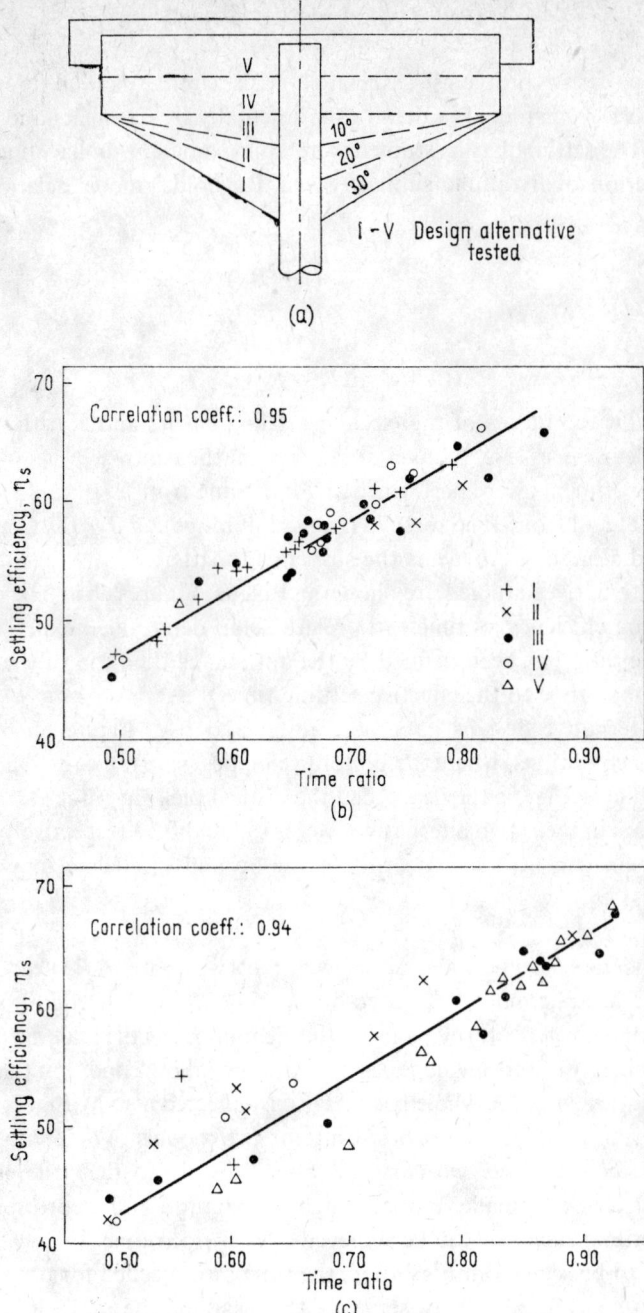

Figs 6(a)–(c). Alternative designs of radial-flow settling tank tested in models.

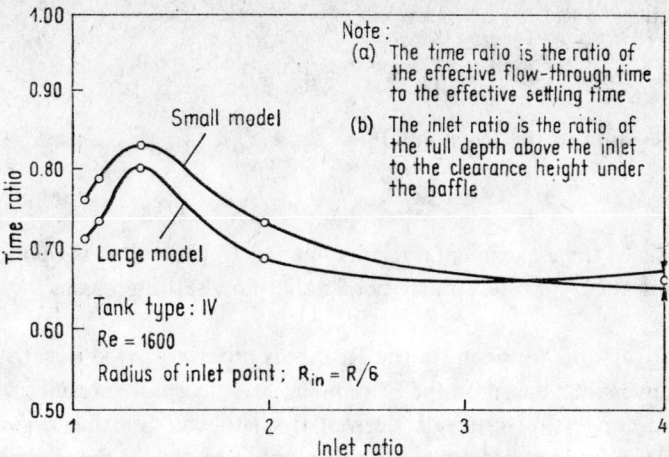

Fig. 7. Time ratio vs inlet ratio in different scales

rate to the settling velocity must assume identical values in the model and the prototype.

It should be noted this conclusion is essentially consistent with that obtained in Thompson's investigations by Hazen's criteria expressed as in Eq. (37). This is logical, since the Ha number includes the velocity ratio v/w that is proportionate to the time ratio.

The pronounced influence of the inlet conditions raises similar ideas as the conclusions related to density currents. The conclusion applying to the geometry of the entire basin (under point c) is believed to contain implicitly Hazen's conventional concept of settling.

On the other hand, if the authors have found a specific range of Re numbers ($Re > 2500$) preferable it is in agreement with the remarks made in connection with self-modelling ranges.

Important contributions to the problems of similarity in radial flow settling basins have been made by Villemonte and Rohlich [37]. They used geometrically similar models corresponding to the scale ratio $\lambda = 4$, with the objective of studying the scale-up criteria by flow-through experiments carried out by dye tracing. The experiments involving different surface loads and basin depth to diameter ratios were evaluated for the typical points on the flow-through hydrograph plotted in a dimensionless system of coordinates.

Since earlier sources have assigned equally important roles to the Fr and Re criteria in modelling hydraulic conditions, Villemonte and Rohlich have checked both experimentally in specific $T_s = Q/F$ surface load ranges. The scale factors of T_s are ($\lambda = 4$).

According to the Froude Law

$$\lambda_{T_s} = \frac{\lambda_Q}{\lambda_F} = \lambda^{1/2} = 2 \tag{42}$$

while according to the Reynolds Law

$$\lambda_{T_s} = \lambda^{-1} = 1/4. \tag{43}$$

Analysis of the experimental results has led to the following conclusions concerning the hydraulic similarity of radial-flow settling basins:

(a) Conversions founded on the Reynolds criteria proved unsatisfactory.
(b) Conversions based on the Fr number gave acceptable results.
(c) The conversion criterion derived from the relationship defining the surface load yielded results even more accurate than those obtained by applying the Fr number.

The validity of Hazen's criteria is again demonstrated by the last conclusion.

Additional details of model studies on radial-flow settling basins have been studied by Albrecht [38], Dague and Baumann [39]. Burdych [40] has reproduced radial flow settling basins on the basis of the Froude criterion ($\lambda = 5$), imposing as a special condition that identical types of flow should prevail in the model and the prototype.

More recently an interesting series of tests related to the hydraulic modelling of a circular settling basin combined with a surge tank has been reported by Tessendorf [41]. Conversion was based on the scale factor $\lambda_v = 1$, which leads in geometrically similar systems and in the case of $\lambda_w = 1$ to the relationship expressed by Eq. (39), and can be traced back to Hazen's criterion. The validity of the conversion criteria was verified experimentally ($\lambda = 6$; $\lambda_{Re} = \lambda = 6$). The fact that in this particular case turbulent flow conditions prevailed in the boundary layer on the basin bottom was considered further evidence for the applicability of the scale ratio $\lambda_v = 1$.

The paper by Christie and Harbinson [42] dealing with model studies on circular settling basins was published almost concurrently with the Hungarian version of this book, contributing additional information on the subject of modelling. Based on earlier research they arrived at the following conclusions concerning the experimental work

(a) In particular cases screened sewage should be used in the model tests.

(b) The model equipment should be designed and operated with Hazen's invariant as the basis of conversion. This means that the velocities and surface loads are equal at the corresponding points in the different systems, thus $\lambda_v = \lambda_{T_s} = 1$. Moreover, the flow should be of the same type (e.g. turbulent) in the model and prototype. To achieve this, sufficiently high Re numbers must be realized in the model as well.

(c) The model should be as large as possible to minimize the scale effect.

From the studies in two models scaled down to 1 : 6 and 1 : 30 Christie and Harbinson eventually established that modelling according to the criterion $\lambda_v = 1$ had failed in this particular case to produce the desired accuracy in removal efficiency. The role of the retention time proved also important and impossible to reproduce properly by observing Hazen's concept.

Concluding the subject of radial-flow settling basins reference is made to a few additional publications, first of all to the papers of Hubbel [43], Porteus [44] and White and Greenless [45].

Vertical-flow settling tanks

The modelling criteria of Dortmund tanks have been investigated by Vágás [46] starting from hydraulic considerations. His objective was to check by experiment the validity of the model law serving as the basis of conversion. The method involved the determination of the flow-through hydrograph by tracer injection. The problem was to correlate the scale factors of velocities, flow rates and characteristic dimensions by producing possibly similar flow-through hydrographs, i.e. similar flow conditions, in two geometrically similar Dortmund tanks of different size ($\lambda = 4$). Specifically, the validity of the expression

$$\lambda_Q = \lambda_Q^\alpha \qquad (44)$$

was checked, just as in the case of Eq. (30b), with the aim of determining which of the laws of hydraulics would correspond to the exponent α_Q obtained. In this way it was actually possible to decide by experiments whether the Fr or Re number is to be used in converting the flow rates. This approach was based on the assumption that corresponding (invariant) velocity distributions result in similar flow-through curves in the two systems. In other words, creating similar velocity distributions in geometrically similar systems, one can expect the flow-through characteristics (flow-through wave and curve) to also be similar. From a large number of tests and experiments the value $\alpha_Q = 2.476$ was derived, supporting with fairly good approximation the validity of Froude's criterion. The scale factors of the other variables have been compared in Table 1.

Table 1

Experimental verification of the Froude criterion for Dortmund-type settling basins

Variable	Symbol	Scale factor		Difference, percentage of Froude number
		by the Froude criterion	by experiment	
Linear dimensions	λ	$\lambda^{1.0} = 4$	$\lambda^{1.000} = 4.000$	0.00
Times	λ_t	$\lambda^{0.5} = 2$	$\lambda^{0.524} = 2.068$	3.40
Velocities	λ_v	$\lambda^{0.5} = 2$	$\lambda^{0.476} = 1.934$	3.30
Discharges	λ_Q	$\lambda^{2.5} = 32$	$\lambda^{2.476} = 30.95$	3.25

Figures 8a–d show (after Vágás) that the flow-through curves determined in two geometrically similar settling tanks of different size for particular loading conditions are approximately similar in shape. Here $t/t_a =$ = time/average retention time; $Q/Q_v =$ labelled discharge/total discharge.

Drawing on the results of Vágás, the Froude criterion has also been adopted by Öllős in a hydraulic study on Dortmund-type settling tanks with mixing chamber [47].

Shifrin [48], from studies on scale-up problems related to vertical flow settling tanks, established the existence of a self-modelling range of Reynolds numbers. He has operated the model ($\lambda = 15$) at Re numbers from 1700 to 50, and remained within the self-modelling range without passing below the lower limit. The results of the model test compared favourably with those obtained in the prototype (using clear water). The hydraulics of vertical flow tanks were analysed in detail by Burdych [49], also including the problems of modelling. The Reynolds numbers $[Re = (vR)/v]$ were calculated from the lowest and highest flow-through velocities in the tank for the two systems of different size. In the case studied ($T_s = 1.4$ m/h) the flow in the prototype (the larger system) was found to be in the laminar and transition ranges while laminar conditions prevailed in the model. It was concluded that the criterion expressed in terms of the Re number (or Re_{crit}) is not as unambiguous as in pipe flow, where $Re_{crit} = 2320$.

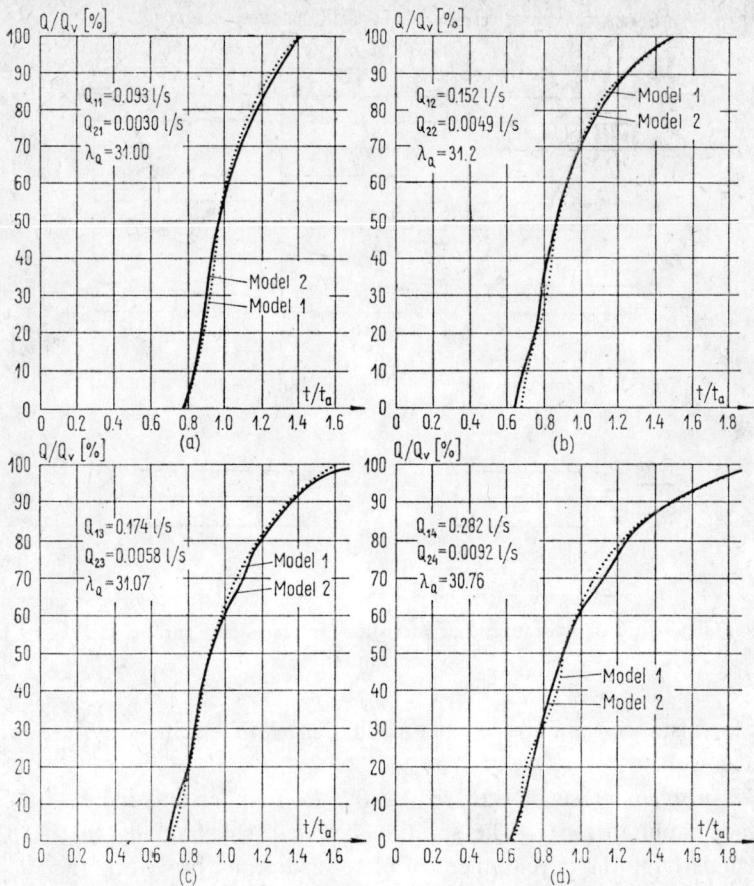

Figs 8(a)-(d). Hydraulic comparison of Dortmund tanks scaled 1 : 4

Plate separators

Any sedimentation device is known to operate at highest efficiency under laminar flow conditions. To attain these the Reynolds number—the ratio of inertial to friction at forces—must be kept as low as possible. On the other hand, the stability of flow is characterized by the Froude number, the ratio of the inertial to gravitational forces. At higher *Fr* numbers the inertial force increases over the gravitational force, indicating more stable flow conditions. This can easily be seen by remembering the currents induced by different densities, where gravity is the dominant force. If the role of the gravitational force becomes less important, the effect of secondary currents

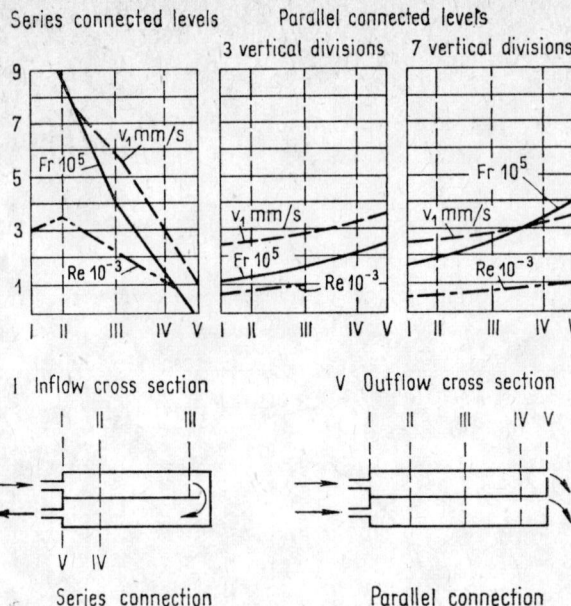

Fig. 9. Comparison of hydraulic characteristics for series and parallel connected plate separators

—as for instance the currents due to differences in specific gravity—diminishes as well.

The problem arose how to reduce the *Re* number and increase the *Fr* number simultaneously. The solution has led substantially to the basic idea underlying the development of plate separators. In writing the *Re* and *Fr* numbers, the hydraulic radius, *R*, of the separator is adopted as the characteristic length

$$Re = \frac{vR}{v} \quad Fr = \frac{v^2}{gR}.$$

In the case of given dimensions these expressions can be rewritten as [50]

$$R = \frac{Rev}{v} = \frac{v^2}{Frg}. \qquad (45)$$

According to Fischerström [51] laminar, stable flow is likely to occur in settling tanks at $Re < 500$ and $Fr > 10^{-5}$. With these introduced into Eq. (36) the values of *R*, or tank dimensions, obtained depart radically from those commonly adopted in practice. Evidently, these criteria cannot be observed, unless the tanks are subdivided into sectors by the installation of suitable baffles. Note that the *Fr* number varies directly with velocity and inversely with the hydraulic radius. However, higher flow velocities entail adverse consequences as well, in that the *Re* number increases and thus turbulence becomes more intensive as a consequence. The only solution consists of reducing the hydraulic radius. In this way the *Fr* number is increased, and the *Re* number reduced simultaneously, both with beneficial effects. The value of *R* can be increased at a given cross-sectional area by increasing the length of the wetted perimeter *K*. The hydraulic problem is thus solved by subdividing the flow space with the help of baffles and diaphragms.

Fischerström [51] has performed numerous experiments on plate separators, some to study new tank designs, others with the aim of converting conventional settling tanks into plate separators. Figure 9 shows one of the alternatives studied in these experiments, in which plate separators of series and parallel connected design were also examined. The variations of the velocities measured in the cross-sections I–V and the *Re* and *Fr* numbers calculated have been plotted in the figures. Parallel connected designs have proved to be superior in general.

Generalization of scale-up methods

Repeated efforts have been made in developing methods which can be extended more or less successfully to different types of settling structures.

The methods using data series from laboratory sedimentation prototype design experiments represent perhaps the simplest approach. The procedure suggested by Eckenfelder and O'Connor [52] for use mainly in wastewater treatment technology is well suited to the case of settling floccular suspensions of low concentration. The experimental equipment consisted of the settling cylinder shown in Fig. 10a. For sampling purposes taps were provided at successive heights. The efficiencies pertaining to the successive times t and the different sludge levels plot as curves (Fig. 10b), while for granular materials straight lines would be obtained. The tangents to these curves at particular points represent the settling velocity.

This line of reasoning has been perfected by Conway and Edwards [53], who have developed a simplified method of correlating the data. Such methods are, in general, poorly founded on similarity theory and are used as aids in solving specific problems encountered in practice. Specifications, such as the scale-up factors of 1.25–1.75 and 1.50–2.0 suggested for surface loading and retention times respectively [52], are similarly theoretically unfounded and apply to special conditions only. Evidently the scale-up factors depend on λ.

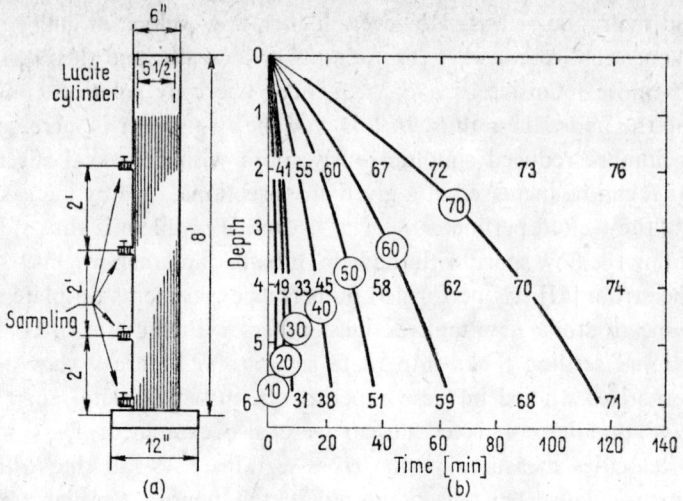

Figs 10(a) and *(b)*. Laboratory evaluation of settling data

The methods involving dimensionless groups and the relationships between them have been derived, on the other hand, from theoretical considerations. Camp [54] studied the influence of flow stability at different hydraulic efficiencies. The Froude number, $Fr = v^2/(gR)$, is known to be a measure of stability. Camp found the hydraulic efficiency to increase together with the Fr number (greater stability). This is clearly shown by Fig. 11, from which one can also infer the existence of a self-modelling range of Fr numbers, where the influence of the latter on the hydraulic efficiency is no more appreciable [the curve section inclining towards the horizontal, $Fr = v^2/(gR) > 10^{-1}$]. Camp, in his investigations, neglected the role of the Re

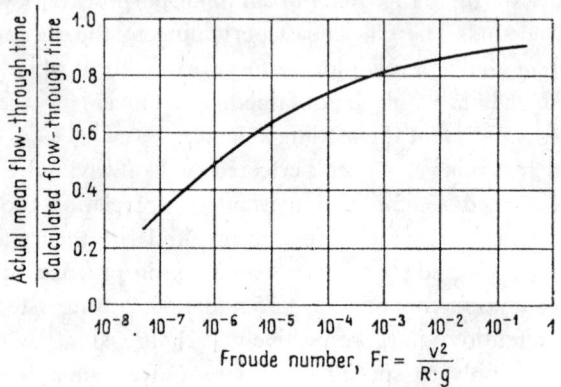

Fig. 11. Hydraulic efficiency of settling tanks vs the Froude number

number. Disregarding a few exceptions, comparable results have been obtained by Schmidt-Bregas [55]. Groche attached critical comments to the conclusions of Camp and Schmidt-Bregas and emphasized that the Re number varies together with the Fr number, which should not be neglected [56]. This in turn implies that the hydraulic efficiency is not as closely related to the Fr number, as would follow from the work of the above authors. Actual conditions are believed to be more closely approximated by relating the hydraulic efficiency to the Fr and Re numbers simultaneously. This dimensionless expression may serve also as the basis of scale-up conversions [57].

Kun Li [58], starting from dimensional analysis, has suggested a method of describing the processes in settling structures. The general form of the governing dimensionless relationship is

$$\frac{\varrho}{\varrho_1-\varrho}\frac{v^2}{gd}=f\left(\frac{dv}{\eta/\varrho_1}\,;\,\frac{Lv}{\eta/\varrho}\right) \tag{46}$$

in which the density simplex, the Fr number, as well as the Re numbers of the particle diameter d and the basin dimension L are involved.

From studies into the engineering and modelling problems associated with the settling of industrial wastewaters, Bramer and Hoak [59] have developed a novel approach, introducing the method widely used in chemical process engineering for scaling-up mass and heat transfer phenomena to the modelling of settling basins. This consists essentially of formulating—possibly in dimensionless form—the empirical relationship of the variables controlling the process and thereby deriving the scale-up equations. The method lends itself fairly easily to generalization, obviously in the validity range of the empirical relationship only.

From the studies on effluents from the steel industry and from data reported in the literature on petroleum refinery effluents, Bramer and Hoak derived, by regression analysis, the following empirical relationship

$$SI=\frac{1.08\times10^{-3}L^{2.09}B^{2.90}H^{0.794}}{Q^{1.13}R^{3.39}} \tag{47}$$

where SI is the sedimentation index, a quantity proportionate to the settling efficiency (min); L, the average settling distance (feet); H, the average basin depth (feet); B, the basin width at the outflow section (feet); R, the hydraulic radius in the outflow section (feet); and Q, the discharge (ft³/min).

Series of data on both longitudinal and radial flow settling tanks were processed to derive Eq. (47). These approaches can be further refined by writing the governing equation in dimensionless form.

Starting from Kun Li's concept [58], the following dimensionless power product was used

$$\frac{\varrho}{\varrho_1 - \varrho} \frac{v^2}{gd} = \text{constant} \left(\frac{dv}{\eta/\varrho_1}\right)^a \left(\frac{Rv}{\eta/\varrho}\right)^b \left(\frac{L}{H}\right)^c \qquad (48)$$

where d is the diameter of the equivalent sphere for the smallest particle to be removed by sedimentation; and v, the velocity of flow.

Dividing the settling process into three typical ranges

(a) constant $= 0.0324$, $a = 0.70$, $b = -0.344$, $c = 2.55$

$$\text{if} \qquad \frac{Rv}{\eta/\varrho} > 525 \quad \text{and} \quad \frac{dv}{\eta/\varrho_1} > 3.00 \qquad (48a)$$

(b) constant $= 0.0401$, $a = 0.517$, $b = -0.114$, $c = 1.36$

$$\text{if} \qquad \frac{Rv}{\eta/\varrho} > 525 \quad \text{and} \quad \frac{dv}{\eta/\varrho_1} < 3.00 \qquad (48b)$$

(c) constant $= 3.61 \times 10^{-8}$, $a = 0$, $b = 2.11$, $c = 1.15$

$$\text{if} \qquad \frac{Rv}{\eta/\varrho} < 525 \quad \text{and} \quad \frac{dv}{\eta/\varrho_1} < 3.00. \qquad (48c)$$

The above formulae were further checked against the data of Rohlich [60] on oil separators. These fitted with acceptable accuracy to Eqs (47) and (48a), but failed to do so to Eqs (48b, c). The settling process of substances having a specific gravity lower than water was described by the following version of Eq. (48) derived by processing a number of data:
constant $= 0.108$, $a = 1.52$, $b = -0.114$, $c = 1.36$

$$\text{if} \qquad \frac{dv}{\eta/\varrho_1} < 3.00. \qquad (48d)$$

A rectangular settling basin was reproduced in a series of geometrically similar models built to the scale factors $\lambda = 4$, 16 and 24, with the aim of verifying the correctness of the foregoing dimensioning formulae and the scale-up method based on them. Except for Eq. (48b), all proved useful. In a subsequent paper Bramer and Hoak demonstrated that the constant k involved in the power product, Eq. (48b), is actually not a constant, but a function of $\bar{R}$ as shown in Fig. 12 [61]. Under these conditions, the exponents of Eq. (48) become

$$a = 0.11, \quad b = 0.64, \quad c = 1.17. \qquad (48e)$$

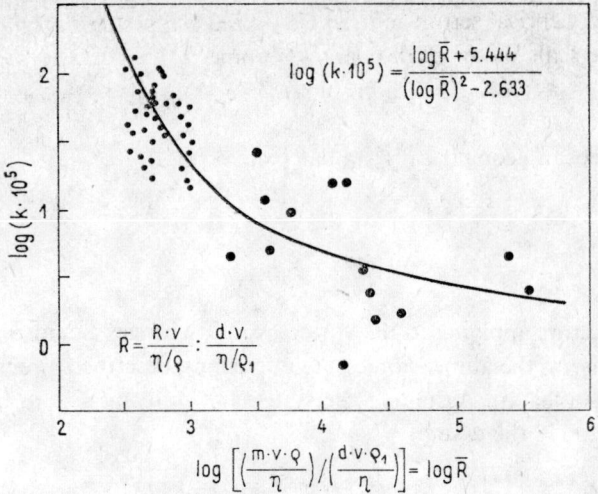

Fig. 12. The constant k vs the *Re* number

The series of model experiments at different scales has also demonstrated the role of the scale effect. Thus it was found impossible to simulate in a certain discharge range the $\lambda=16$ model by that corresponding to $\lambda=24$. Similar experiences occurred with the related systems of $\lambda=4$ and $\lambda=16$. The smallest model size by which the prototype structure could still be reproduced was characterized by the limit values $Re=2100$, and

$$Re_R = \frac{Rv}{\eta/\varrho} = \frac{Re}{4} = 525 \qquad (49)$$

which clearly indicates the transition from laminar to turbulent conditions. The validity ranges of the expressions in Eqs (48a–c) have been obtained from this.

The scale factors applied in modelling the discharge Q, the feed rate q of suspended solids and the time t were

$$\lambda_Q = \lambda^x, \quad \lambda_q = \lambda, \quad \lambda_t = \lambda^2 \qquad (50a\text{--}c)$$

where the magnitude of x is found from the corresponding empirical expression. It should be noted also that the surface profiles of the sediments deposited were also compared in the different models. The method proved in general satisfactory.

Continuing the line of reasoning adopted by Bramer and Hoak [61], based on their relationships, the criterion of scaling-up will be formulated subsequently for some of the cases commonly encountered. The expressions will be written in a form allowing them to be compared with the methods applied by other authors.

Assuming identical settling efficiencies in the two systems of different sizes ($\lambda_{SI}=1$), the scale factor of discharges becomes

$$\lambda_Q = \lambda_L^{1.85} \lambda_B^{2.56} \lambda_H^{0.70} \lambda_R^{-3.0} \tag{51}$$

which reduces in geometrically similar systems to

$$\lambda_Q = \lambda^{2.11} \tag{52}$$

whence

$$\lambda_v = \lambda^{0.11}. \tag{53}$$

The scale factors applying to the various validity ranges can also be derived on the basis of the dimensionless relationships described previously and given as Eqs (48a–d). Assuming the systems of different size to be geometrically similar, in the case of

$$\lambda_\varrho = \lambda_{\varrho_1} = \lambda_\eta = \lambda_g = \lambda_d = 1$$

Eqs (48a–d) yield the exponents of the expressions

$$\lambda_v = \lambda^{\alpha_v}, \quad \lambda_Q = \lambda^{\alpha_Q} = \lambda_v \lambda^2 \tag{54a-b}$$

as $\alpha_v = -0.18, \quad -0.07, \quad -19.2, \quad -0.19$

$\alpha_Q = 1.82, \quad 1.93, \quad -17.2, \quad 1.81$

where $\alpha_Q = \alpha_v + 2$.

From the above the following can be concluded:

(a) All calculated α_v and α_Q values (with a single exception) are scattered around $\alpha_v \approx 0.00$ and $\alpha_Q \approx 2.0$. The outstanding value obtained for Eq. (48c) is interpreted as a sign of questionable validity of the expression. Conversely the fair agreement of the remaining data reflects remarkable consistency.

(b) Comparing the findings of several authors it is noted that the results, although obtained by totally different methods of processing, are usually in good agreement with the scale factors found on the basis of the surface loading and Hazen's criterion. Equations (37) or (38) and (39) yield, in fact, $\alpha_v = 0.0$ and $\alpha_Q = 2.0$ under these particular conditions ($\lambda_d = \lambda_w = 1$). The same follows from the criterion of Eq. (28). It should be emphasized that results obtained independently by different authors have been analysed and compared.

(c) Unfortunately, no explanation is offered in connection with Eqs (50b, c). The practical value is therefore considered questionable since they appear to be inconsistent. [For example, for $\lambda_t = \lambda^2$ it would follow that $\lambda_Q = \lambda$, contrasting with Eq. (50a).]

Another potential approach essentially consists of adopting an experimental modelling technique so as to have unit conversion factors (similarity transformation parameters) for some of the relevant physical variables, such as the retention time, surface loading, etc. In practical terms this means that the familiar technique of "trivial modelling" is used which can be applied successfully in model studies on a number of technological processes. With reference to earlier publications [62, 63] a theoretical example will be given.

Consider as a starting point the fundamental dimensioning formula. The average retention time is

$$t = V_s/Q_w$$

where t is the theoretical detention time; V_s, the volume of the settling basin; and Q_w, the flow rate through the basin. The surface loading is defined by the second basic formula

$$T_s = Q_w/F$$

where F is the water surface area in the basin.

Introducing the similarity transformation parameters, the above expressions lead to

$$\lambda_t = \frac{\lambda_{V_s}}{\lambda_{Q_w}} \tag{55}$$

and

$$\lambda_{T_s} = \frac{\lambda_{Q_w}}{\lambda_F}. \tag{56}$$

Combining Eqs (55) and (56) to eliminate λ_{Q_w}

$$\lambda_t = \frac{\lambda_{V_s}}{\lambda_{T_s}\lambda_F}, \quad \lambda_{T_s} = \frac{\lambda_{V_s}}{\lambda_t\lambda_F}. \tag{57a, b}$$

In geometrically similar systems

$$\lambda_F = \lambda^2, \quad \lambda_{V_s} = \lambda^3. \tag{58a, b}$$

If the transformations expressed by Eqs (58a, b) are taken into account, Eqs (55) and (57) can be rewritten as follows:

$$\lambda_t = \frac{\lambda^3}{\lambda_{Q_w}}, \tag{59}$$

$$\lambda_{T_s} = \frac{\lambda_{Q_w}}{\lambda^2}, \tag{60}$$

$$\lambda_t = \frac{\lambda}{\lambda_{T_s}}. \tag{61}$$

These three expressions represent the relationships between the scale factors of the major variables involved in the designing of settling basins. From Eqs (59)–(61) the following conclusions of practical interest can be reached:

(a) In order to ensure the same detention times in the model (double prime) and the prototype (single prime) the transformation

$$\lambda_t = \frac{t'}{t''} = 1 \tag{62}$$

leads, according to Eq. (57), to the expression

$$\lambda_{Q_w} = \frac{Q'_w}{Q''_w} = \lambda^3. \tag{63}$$

For instance, the prototype discharge Q'_w is obtained by multiplying the model discharge Q''_w with λ^3.

(b) According to Eq. (60) the same surface loading ($\lambda_{T_s} = T'_s/T''_s = 1$) will result in the prototype and the model if the condition

$$\lambda_{Q_w} = \frac{Q'_w}{Q''_w} = \lambda^2 \tag{64}$$

is satisfied. Evidently, in the case of a particular experimental design alternative Eqs (63) and (64) are impossible to satisfy simultaneously. Moreover, this is expressed also by Eq. (61); in the case of $\lambda_t = \lambda_{T_s} = 1$ one has $\lambda = 1$. Consequently, the condition $\lambda_t = \lambda_{T_s} = 1$ cannot be satisfied, unless the scale is 1 : 1.

(c) As can be seen from Eq. (61), the condition $\lambda_t = 1$ can also be satisfied according to

$$\lambda_{T_s} = \lambda \tag{65}$$

substantially conveying a statement identical to that expressed by Eq. (64). (No more than a simplification by the scale factor related to surface is involved.)

From the theoretical considerations outlined it can be realized that in modelling and dimensioning settling structures, either the detention time or the surface loading must be adopted as the basis of calculations. In practice, this problem is commonly solved by taking the settling characteristics of the suspended substance into consideration. In the case of granular materials, where the size or the settling velocity of the individual particles remain virtually unchanged during the settling process, the surface loading is preferably adopted as the basis of dimensioning. In contrast, in the case of flocculating or coagulating settling particles the detention time is the parameter controlling the design.

The theoretical results of interest to the practical designer engaged in dimensioning on the basis of the average detention time can be summarized as follows:

(a) The model should possibly be geometrically similar (or approximately similar) to the prototype. In this way the flow pattern and conditions can be more accurately reproduced.

(b) In operating the model and in dimensioning it is advisable to observe the criterion $\lambda_t = 1$, and assuming in this case an aggregated, coagulating material, to dimension the structure on the basis of the retention time.

(c) Consequently from above, the discharges in the prototype and the model (the sewage inflow to the structure, the return flow, etc.) can be converted to each other by Eq. (63) once the scale factor, or scale ratio is known.

It is finally deemed necessary to attach a few remarks to the conclusions under (a). In the consideration of similarity theory attention has been focused, logically, on the practical requirements of dimensioning, rather than on achieving hydrodynamic similarity. The approach can be justified as follows.

In open settling structures hydraulic similarity is known to be controlled by the Froude criterion, according to which $\lambda_t = \lambda^{1/2}$. This, however, is incompatible with Eqs (63) and (64), since for $\lambda_t = 1$, $\lambda = 1$ would also result. The relationship written on the basis of the Froude number could be incorporated in Eqs (59)–(61), from which the familiar expressions

$$\lambda_{Q_w} = \lambda^{5/2} \quad \lambda_{T_s} = \lambda^{1/2} \quad \lambda_t = \lambda^{1/2} \qquad (66\text{a--c})$$

would follow. It should be realized, however, that Eqs (66a–c) obtained by assuming the validity of the Froude Law are mainly used as modelling relationships in studying flow conditions and analysing flow patterns. In the case considered, the use of Eq. (62) is more appropriate, which implies at the same time, in good agreement with experimental evidence, that in the settling process taking place in the model and prototype the corresponding times t' and t'' are identical, if the settling properties of the particles are also identical. Consequently, if a retention time of 1 h is to be achieved in the prototype, then the model experiment must also be operated with a retention time of 1 h.

It should be noted that measured by the exact theorems of similarity theory, the solution described is no more than an approximation, but its applicability has been demonstrated in wastewater treatment practice. Complete similarity of settling structures is impossible to attain, as can be shown analytically.

The effect of turbulence

The effect of turbulence has already been mentioned repeatedly. Attention is directed again at Eqs (22), (25), (26) and (28b) [16, 64, 65]. It is important to remember that the same type of flow must prevail in the model and prototype. For this purpose the Re_{crit} value, indicating the transition from laminar to turbulent conditions, is significant. A range $Re_{crit} = (vR)/v = 500$–2000 is mentioned in the relevant literature [49, 59]. As pointed out previously, the Re number is not as positive a criterion in the case of tanks as in pipe hydraulics, since other factors, such as the inlet arrangement, the design of structural elements, etc., also play important roles. For example, the hydraulic studies by Mau [66] on the inlet device of a horizontal-flow settling structure have shown measurement results to be little affected by varying the Re number, provided it remained greater than 200.

Pikalov [67], in his studies on the modelling problems related to sediment-carrying water, has assumed that in turbulent flow one of the similarity criteria is expressed by the ratio $\bar{v}/u^*$ (where $\bar{v}$ is the flow velocity averaged over time, the average velocity; and u^* is the vertical pulsational component of velocity)

$$\left(\frac{\bar{v}}{u^*}\right)' = \left(\frac{\bar{v}}{u^*}\right)''. \tag{67}$$

Pikalov assumed the similarity criterion of the movement of suspended sediment particles in turbulent flow to be

$$\left(\frac{w}{u^*}\right)' = \left(\frac{w}{u^*}\right)''. \tag{68}$$

Combining the above expressions [68]

$$\left(\frac{\bar{v}}{w}\right)' = \left(\frac{\bar{v}}{w}\right)'' \tag{69}$$

where, assuming kinematic similarity, the average velocity $\bar{v}$ can be replaced by the mean velocity v_m. Hence one obtains

$$\lambda_v = \lambda_w. \tag{70}$$

If the validity of the Froude criterion is accepted, then $\lambda_w = \lambda^{1/2}$. An additional important similarity criterion is that $\lambda_C = 1$, according to which the concentrations C of the suspended sediment at the corresponding points of the model and prototype should be equal.

On the strength of experimental evidence, Ivicsics [68, 69] suggested model designs in which $\lambda_{w_t} = 1$, so that $w'_t = w''_t$ (where w_t is the settling velocity in turbulent flow). The scale factor of basin length is from Eq. (28a)

$$\lambda_l = \frac{\lambda_v}{\lambda_{w_t}} \lambda_h = \lambda_v \lambda_h \tag{71}$$

by observing the Froude criterion

$$\lambda_h = \frac{\lambda_l}{\lambda_v} = \lambda_l^{1/2}. \tag{72}$$

The scale factor of discharges is

$$\lambda_Q = \lambda_l \lambda_h \lambda_v = \lambda_l^2. \tag{73}$$

It should be noted that Eq. (73) is identical with Eqs (38a) and (64), though under conditions which differ in part from the above, and agrees with fair approximation with Eq. (52).

As pointed out further by Ivicsics, the criterion $\lambda_d = 1$ does not guarantee compliance with the condition $\lambda_w = 1$. Starting, for instance, in the case of turbulent flow from the criterion $\lambda_d = 1$, it is impossible to reach $\lambda_w = 1$, since the flow velocity v' diminishes to v''. From Fig. 13 it can be seen that $w'_t \neq w''_t$ and $w' - w'_1 \neq w'' - w''_1$, since $w'_t = w - w'_1$, $w''_t = w'' - w''_1$ and $w'_1 \neq w''_1$, where w_1 denotes the decrease in settling velocity (over the value w in

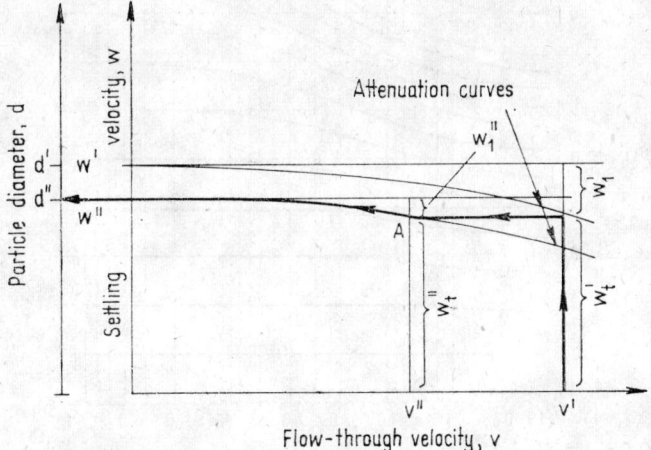

Fig. 13. Description by attenuation curves

stationary water) due to turbulence. The velocity reduction curves in the figure indicate the effect of turbulence. From these curves the d'' and w'' values to be used in the model experiments are found graphically for the case $w'_t = w''_t$. (The graphical construction procedure is indicated by arrows). In problems of practical interest different particle sizes are involved, which should be evaluated according to Pikalov by resolving them into fractions, e.g. in satisfying the condition $w'_t = w''_t$.

In deriving the scale-up criteria for sedimentation basins with turbulent flow the Dobbins–Camp approach [70] is also considered a potential starting basis. Assuming the velocity distribution in rectangular settling basins to be a parabolic one, Camp has derived in terms of the shear stress τ, the density ϱ and the depth H the following expression

$$\varepsilon = 0.075 H \sqrt{\left(\frac{\tau}{\varrho}\right)} = 0.075 H u_* \qquad (74)$$

where ε is a coefficient representing the intensity of turbulent mixing (diffusion), and u_*, the mean shear velocity.

To take the effect of turbulent flow into consideration Camp plotted (Fig. 14) the efficiency η_s of settling against the quantity $(wH)/2\varepsilon$ (a dimensionless number representing the intensity of turbulent diffusion) obtained as

$$\frac{wH}{2\varepsilon} \approx 122 \frac{w}{v} = 122 \frac{w_t}{v} \frac{w}{w_t}. \qquad (75)$$

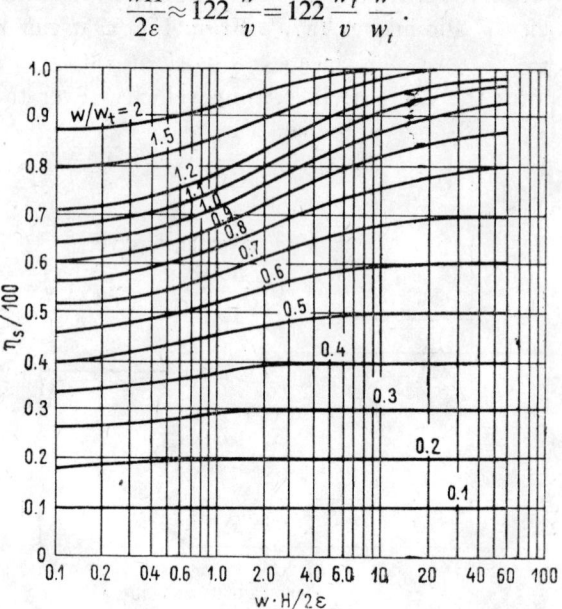

Fig. 14. Allowance for turbulent flow in rectangular settling tanks

The ratio w/w_t is used as the parameter. According to Hazen's theory the ratio w/T_s can be substituted for w/w_t. Given the values of the variables w and v, the magnitude of w_t is found from the figure and the effective length of the settling basin therefore is $L=(Hv)/w_t$.

Another potential approach in deriving the model–prototype relationship is based on the consideration that in order to obtain identical settling efficiencies in the two systems, the dimensionless groups $(wH)/\varepsilon$ and w/v must be invariant

$$\lambda_v = \lambda_w = \lambda_{w_t} = \lambda_{T_s} \tag{76}$$

and

$$\lambda_\varepsilon = \lambda_w \lambda_h. \tag{77a}$$

Specifying the criterion $\lambda_w = 1$, one has

$$\lambda_\varepsilon = \lambda_h. \tag{77b}$$

In undistorted models

$$\lambda_l = \lambda_h \frac{\lambda_v}{\lambda_{w_t}} = \lambda_h \tag{78}$$

and the scale factor of discharges ($\lambda_w = \lambda_{T_s} = 1$) becomes

$$\lambda_Q = \lambda_l^2 \tag{79}$$

conforming with the modelling criteria of Eqs (38a), (64) and (74).

Finally, it should be noted that the dimensionless number

$$\frac{H}{\varepsilon} \sqrt{\left(\frac{\tau}{\varrho}\right)} = \frac{Hu_*}{\varepsilon} \tag{80}$$

derived from Eq. (74) and that obtained from Eq. (75) as $(wH)/\varepsilon$—which form the basis of scaling-up in this case—may be regarded substantially as the modified forms of the *Pe* number defined for component transfer (if the mixing coefficient ε is regarded as a quantity analogous to the diffusion constant D).

Camp [70] in his paper, which might now be called a standard, experimentally demonstrated the approximate validity (within certain limits) of Froude's criterion in modelling the flow conditions in settling basins. He specified geometric similarity as one of the essential conditions. It seems necessary, however, to underline the discussion of Eliassen [71], who emphasized that friction plays a role comparable in importance to gravity. He suggested that the gravitational force predominates mainly around the inlet part of settling basins, where relatively higher velocities prevail.

Comparison of scaling-up criteria for settling basins

First of all it must be realized from the literature on the subject that the model laws and similarity criteria applied to the entire structure or their discrete component parts, or to different operating ranges of various types of settling basin, have failed to produce consistent results. Neither can any single comprehensive modelling method (conversion procedure) be expected to solve this complicated, ramified problem. The primary reason for this is believed to be related to the complexity of the phenomena.

Nevertheless, from the publications reviewed and from the results of my investigations, some conclusions of more or less general nature may still be arrived at:

(a) Models intended for studying the hydraulics of settling basins as open flow systems should be scaled on the basis of the Froude criterion. For successful application the flow must be of the same type (transition–transition or turbulent–turbulent) in both the model and the prototype. However, in practice adherence to this criterion is not always possible; the flow may be unstable and observation of Froude's criterion may lead to erroneous conclusions.

(b) Observing specific conditions the conversion relationships derived from Froude's invariant are consistent with those obtained by introducing the Ri number. It should be pointed out, however, that for this agreement to exist the corresponding densities and temperatures must be equal. Model studies on the wind effect, specifically on wind induced currents (a potential problem in large sedimentation basins, just as in storage reservoirs), can be demonstrated to yield fairly acceptable results by keeping the Ri number invariant [72].

(c) To obviate the problem mentioned under (a) several authors have attempted or suggested to adopt the Reynolds criterion instead. As an approximate alternative velocities corresponding to the lower limit of the turbulent range have been applied. But in practical applications these efforts were observed to yield rather inconsistent results. The Reynolds criterion seems better suited to modelling types of settling tank, where the flow radically differs from that in open installations, i.e. in plate or tube separators and some submerged parts of settling tanks, but even here the Fr and Re numbers act simultaneously. It should also be noted that in some settling tank configurations the definite determination of the Re_{crit} value may also present appreciable difficulties.

(d) As a compromise between the *Fr* and *Re* numbers a linear combination of the two has been suggested in the form of the product *FrRe* from which the invariant group

$$FrRe = \frac{v^3}{gv}$$

is obtained, leading to the scale factor $\lambda_v = 1$ (in the case of $\lambda_g = \lambda_v = 1$). In other words, this is a case of the equal velocity or equal surface load scaling method, often mentioned in the subject literature.

(e) The method outlined under (d) is consistent with the Hazen principle (Hazen's invariant). Therefore, in this case the scaling criterion is written as $\lambda_v = \lambda_{T_s} = 1$. The important role of the surface loading should be emphasized in this context. The applicability of this method to the sedimentation of inorganic (discrete) suspended particles has been experimentally demonstrated.

(f) Cases are encountered (e.g. sedimentation of flocculating substances) where the role of detention time predominates over that of the surface load. The *Ha* number given by Eq. (37) is then preferably rewritten ($Q = V/t$) as

$$Ha = \frac{Q}{wLB} = \frac{V}{twLB} = \frac{D}{tw} = St \tag{37a}$$

which may be regarded a special form of Strouhal's invariant. This is at the same time modelling unsteady flow phenomena.

(g) Two different interpretations can be given to the latter equations: if, for instance, the criterion $\lambda_v = \lambda_w = \lambda_{T_s} = 1$ must be observed in order to reproduce the technological process of sedimentation, a result complying with the invariant *FrRe* will be obtained. On the other hand, if adherence to the criterion $\lambda_t = 1$ is considered desirable, Eq. (37a) leads to the relationship in Eq. (63).

The scale factors obtained on the basis of the principal invariants have been summarized in Table 2, involving the dimensional variables time *t*, velocity *v*, and the discharge *Q*, together with some dimensionless quantities, which play essential roles in modelling settling structures. Evidently, the table can be extended to other variables as well.

As will be appreciated, the spectrum of alternative conversion methods is a very broad one from which the choice of the correct or most closely approximating solution is governed by the nature of the particular problem and the initial conditions. To illustrate the broadness of the spectrum it should be recalled that in the relationship $\lambda_Q = \lambda^\alpha$, for instance, the exponent α may assume values 1, 2, 2.5 and 3.

Table 2

Main invariants and scale factors involved in scaling-up settling basins

Scale factor	Condition equation of invariance			
	$Re' = Re''$	$(Fr \cdot Re)' = (Fr \cdot Re)''$ $Ha' = Ha''; v' = v''$	$Fr' = Fr''$	$Ha' = Ha''; t' = t''$
	1	2	3	4
Time $\lambda_t =$	λ^2	λ	$\lambda^{1/2}$	λ^0
Velocity $\lambda_v =$	λ^{-1}	λ^0	$\lambda^{1/2}$	λ
Discharge $\lambda_Q =$	λ	λ^2	$\lambda^{5/2}$	λ^3
Re number $\lambda_{Re} =$	λ^0	λ	$\lambda^{3/2}$	λ^2
$FrRe$ number $\lambda_{FrRe} =$	λ^{-3}	λ^0	$\lambda^{3/2}$	λ^3
Fr number $\lambda_{Fr} =$	λ^{-3}	λ^{-1}	λ^0	λ
Ha number $\lambda_{Ha} =$	λ^0	λ^0	λ^0	λ^0

Note: The columns of the table are arranged in decreasing (e.g. λ_t) and increasing (e.g. λ_v and λ_Q) order of the exponent of λ. Thus in the case of λ_Q the exponents are 1; 2; and 2.5; 3. The last row is an exception, where the exponent is invariably zero.

Regarding the conversion of the dimensionless characteristic quantities, it will be seen from the tabulation that in the case of the criterion $FrRe=$ idem, the scale factor λ_{Re} of the Reynolds number equals the scale factor of lengths λ. For this reason cases are conceivable where rather large models must be constructed even under this criterion in order to reproduce the turbulent flow prevailing. This applies even more so to models based on the Froude Law, or on $\lambda_t = 1$ (where the exponents of λ are 3/2 and 2, respectively).

It is worth noting that the criterion $Ha=$ idem and $St=$ idem are satisfied in each of the four alternatives tabulated.

2.1.4 Filters

No thorough investigation has so far, in my knowledge, been made into the theoretical aspects of similarity when modelling the technological processes of filtration in water and wastewater treatment. However, a number of reports have indicated that filtration has been studied extensively in experiments using devices ranging in size from laboratory models to full-scale plant equipment. The resulting inconsistencies can readily be traced back to this circumstance. The modelling problems associated with deep-bed filters will be reviewed in brief, emphasizing that further detailed theoretical and experimental studies are needed on this subject.

The hydraulic aspects of scaling-up

Model studies into the hydraulics of flow in porous media have been described in several reports and papers, mostly in connection with seepage problems [73, 74]. In Hungary, the scale-up criteria of seepage within the validity range of the Darcy Law, specifically in that of the Kozeny–Carman equation, have been considered by Mosonyi and Kovács [75]. In a later paper [76] a comprehensive review has been presented on some of the similarity criteria involved in seepage phenomena. With reference to these studies, the following dimensionless numbers are suggested as modelling criteria for flow in granular filter media.

In the laminar range of two-phase flow [73, 75]

$$MK = \frac{Fr}{Re} = \frac{vv}{l^2 g} \tag{81a}$$

or [76]

$$Po = \frac{l^2 (dp/dl)}{\lambda v} \propto \frac{l^2 \gamma}{\eta v} = \frac{l^2 g}{vv} \tag{81b}$$

where MK is the invariant group of Mosonyi and Kovács and

$$Po = \frac{1}{MK} = \text{the Poiseuille number.}$$

The characteristic length l is preferably replaced with, for example, the particle diameter d.

In the range of three-phase flow [74, 77]

$$I = \frac{We}{Re} = \frac{v\eta}{\sigma} \tag{82}$$

where I is the capillary (Horváth) number.

The various hydraulic quantities, such as velocity or discharge, can be converted from one system to the other using the expressions given in the literature referred to above.

Dimensionless numbers and relationships describing filtration

In recent years certain advances have taken place in the determination of dimensionless numbers on which the filter models are based. Of these examples the results of Ison and Ives will be dealt with first. By dimensional analysis they derived the following relationship [78]:

$$\Phi\left(\frac{C}{C_0}, \frac{d}{H_f}, \frac{e_p}{d}, \frac{vd}{\eta/\varrho}, \frac{\overline{(\varrho_i-\varrho)}e_p^2 g}{\eta v}, \frac{\varrho_t e_p^2 v}{\eta d}, \frac{vd}{D}\right)=0 \qquad (83)$$

where C is the concentration of suspended matter in general; C_0, the initial concentration of suspended matter; d, the particle size of the filter medium; H_f, the thickness of the filter bed; e_p, the particle size of the suspended matter; v, the characteristic flow velocity; $\overline{(\varrho_i-\varrho)}$, the average density differential between the suspended particles and the medium; and D, the diffusion constant of the flowing medium.

The dimensionless numbers involved in the above expression will be seen to contain, besides the characteristic simplexes, the modified forms and combinations of the Re, Fr and Pe numbers. From several sources it is inferred that of the dimensionless numbers of dynamic character describing filtration, the Fr and Re numbers play the most important roles [76]. Of the dimensionless numbers describing component transport, foreign authors emphasize the importance of the Pe number in the study of deep-bed filters [79]. The processes of coagulation and flocculation are, moreover, characterized (especially in the case of contact filtration) by the Camp number

$$Ca=Gt \propto \frac{vt}{l} \qquad (84)$$

the inclusion of which may also be found necessary in considerations related to filtration [80]. The Camp number may be regarded as the inverse of the homochronous number [76] and plays a dominant role mainly in clarification technology (see Sections 2.2.1 and 2.2.2).

In this doctorial thesis Ison has written, for the deep-bed filters examined, the following dimensionless numbers [79]:

the interception parameter

$$I=\frac{e_p}{d} \qquad (85)$$

the inertial parameter

$$M=\frac{\varrho_t e_p^2 v_f}{9\eta d} \qquad (86)$$

the settling parameter (based on Stokes law)

$$G=\frac{g(\varrho_t-\varrho)e_p^2}{18\eta v_f}=1 \qquad (87)$$

the Re number

$$Re=\frac{dv}{\eta/\varrho} \qquad (88)$$

the Pe number (in modified form)

$$Pe = \frac{3\pi \eta e_p v_f d}{kT} \qquad (89)$$

where k is Boltzmann's constant; T, the absolute temperature; and v_f, the filtration rate.

Reviewing the studies of several authors, a comprehensive analysis of the applicability of dimensionless groups in filter designing was presented by Ives [80, 81], without mentioning the problems of scaling-up. The modelling approach adopted is also, in some instances objectionable on theoretical grounds [82].

The physico-chemical features of filters can also be expressed with the help of dimensionless numbers, as exemplified by the studies of Fitzpatrick and Spielman [83, 84], who expressed the filter coefficient λ_f in terms of the following groups:

the adhesion number

$$N_{AD} = \frac{Qd^2}{9\pi \eta B_e^3 v_f} \qquad (90)$$

the gravitational number

$$N_{GR} = \frac{2(\varrho_l - \varrho)gd^2}{9\eta B_e^3 v_f} \qquad (91)$$

the number involving the zeta-potential

$$N_\xi = \frac{3\varepsilon e_p\, \xi_p \xi_f}{2Q}. \qquad (92)$$

In these dimensionless numbers the following symbols are used: Q is Hamaker's constant; B_e, dimensionless porosity number; ξ_p, ξ_f, the zeta potentials related to the suspended particles and the filter medium, respectively; and ε, the dielectric constant.

In conclusion it is observed that although dimensionless numbers are gaining popularity for describing filtration processes, no scaling-up methods founded adequately on similarity theory are yet available, at least in water and wastewater treatment technology. For this reason the models are commonly designed according to scaling-up principles derived from practical observations.

Experience gained from scaling-up experiments

Studying the filtration characteristics of industrial effluents, Donovan [85] performed parallel experiments with filters of 6 in. and 3 ft diameter with plant-scale equipment, in order to compare the results. He found the

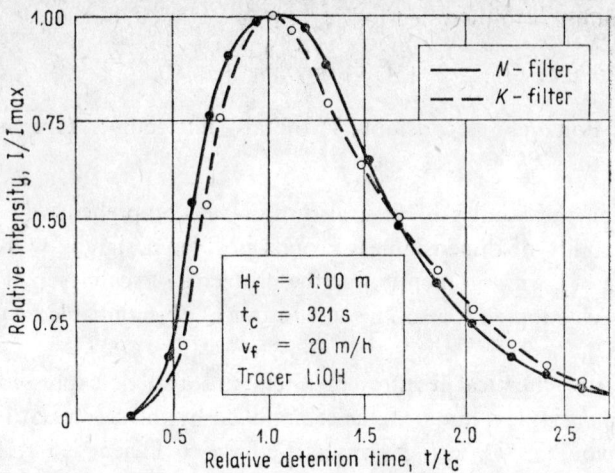

Fig. 15. Dimensionless flow-through hydrographs (closed sand filter)

data obtained from different sizes to be substantially in agreement. Adin and his co-workers studied the problem of applying the filter equations to the dimensioning of pilot plant filter equipment [86].

Investigations on closed rapid filters by Juhász and myself [87, 76] used equipment of two different sizes denoted K and N. The retention times were measured using LiOH as a tracer material; the distributions obtained are shown in Fig. 15. The comparison of the two flow-through hydrographs reveals that although the two are almost identical in character, the dispersion of the tracer is more pronounced in the filter N. The K unit had a higher filter bed depth to column diameter (H_f/D_f) ratio than unit N (1 : 0.25 m = 4 > 1 m : 0.685 = 1.46 m) and accordingly the flow-through hydrograph resembles more closely the one of a so-called tube-type reactor. For technological comparison Fig. 16 is shown, relating the efficiencies of suspended solids removal in the K and N systems. The mean values of the experiments performed with and without chemical feed [$Al_2(SO_4)_3$] would suggest with fair approximation the criterion $\lambda_\eta = 1$, implying the two systems to be equivalent as regards efficiency (assuming identical operating conditions).

From the above the following conclusions of practical interest have been arrived at [76]:

(a) In modelling deep-bed filters the same fluid and filter medium should be used in the model and the prototype.

(b) The dimensions of the equipment should not be distorted in the direction of flow.

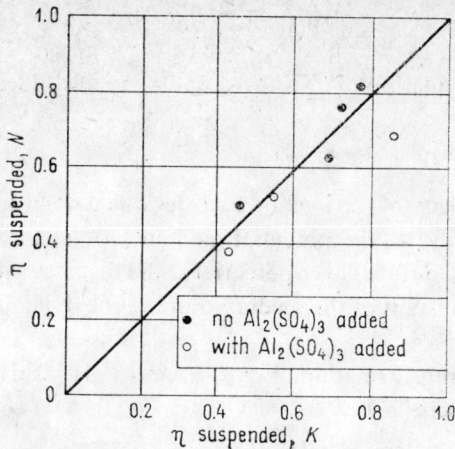

Fig. 16. Comparison of mean filter efficiencies for two rapid filters (N and K) of different size

(c) The dimensions perpendicular to the flow may be reduced to a certain extent, the limit being set by the wall effect.

(d) It is important to ensure identical initial and boundary conditions of equipment and operation.

It should be noted that these considerations were observed in the investigations on the pilot equipments N and K mentioned before.

Modelling cake filtration

The different designs of filters operating on the cake principle (e.g., band filters and filter presses) are used mainly for sludge treatment in water and wastewater technology. For the similarity and modelling problems related to these devices reference is made to the work of Zingler [88].

2.2 CHEMICAL TREATMENT

2.2.1 Flocculators

The zeta-potential and electrophoretic mobility

A colloidal particle moving in an electric field under the influence of the potential gradient is acted upon by the field force and friction. The transport equation written after Sennett and Olivier [89] for the direction of movement is

$$E\omega \, dx = \frac{d^2v}{dx^2} \, dx. \tag{93}$$

The electrostatic equation of Poisson must also be introduced

$$v^2\psi = -\frac{4\pi\omega}{\varepsilon} \tag{94}$$

where v is the velocity of the ions or particles due to field intensity; the electrophoretic mobility; E, the potential gradient; ω, charge related to unit volume; ψ, electric double-layer potential at the distance x from the surface; ε, the dielectric constant of the medium; and η, the dynamic viscosity of the medium.

By similarity transformation, Eqs (93) and (94) yield two relevant dimensionless numbers [76]

$$\pi_1 = \frac{E\omega l^2}{\eta v}, \quad \pi_2 = \frac{\psi\varepsilon}{\omega l^2}. \tag{95a,b}$$

Assuming the product of π_1 and π_2 to be constant in similar systems

$$\frac{\varepsilon E\psi}{v\eta} = \text{constant}, \quad v = \text{constant} \, \frac{\varepsilon E\psi}{\eta}.$$

A particular value of ψ is termed the zeta potential ζ. Thus with the substitution $\psi = \zeta$, further for the case constant $= 1/4\pi$, the familiar Helmholtz-Smoluchowski expression is obtained

$$v = \frac{1}{4}\pi \frac{\varepsilon\zeta E}{\eta}, \quad \zeta = 4\pi \frac{\eta v}{\varepsilon E}. \tag{96a,b}$$

The solution of Eqs (93) and (94) by a similarity theory approach leads substantially to the fundamental relationship of Eq. (96).

The velocity gradient and energy dissipation

The relationship between the velocity gradient and energy dissipation plays—according to the original assumptions of Camp and Stein [90]—a role of decisive importance in coagulation and flocculation processes. The results of subsequent research have demonstrated that a design philosophy based on the concept of the velocity gradient may be successful in describing the hydraulics of other technological equipment as well, which suggests the possibility of applying at least to a certain extent uniform principles in dimensioning work.

The velocity gradient prevailing at different points within the flow and the shear stresses predominating under laminar conditions result partly from the turbulence of the fluid introduced into the structure and partly from the agitation processes induced in the structure itself. Camp and Stein have suggested a relationship between the velocity gradient and the energy dissipated, in terms of the shear stress prevailing under laminar conditions.

The work performed in unit time within unit fluid volume, specifically the energy dissipated in laminar flow, is

$$d = \eta \left(\frac{dv}{dz}\right)^2 = \eta G^2 \qquad (97)$$

with

$$G = \sqrt{\left(\frac{d}{\eta}\right)} = \sqrt{\left(\frac{D}{\eta V}\right)} \qquad (98)$$

where G is the velocity gradient, which has been replaced in finite volume by the value $G = \sqrt{\bar{G}^2}$ averaged over time; $d = D/V$, the energy dissipated in unit time and unit volume of the moving fluid; $D = E/t$, the energy dissipated in unit time within the moving fluid; and V, the fluid volume.

Rufy and his co-workers have studied the role of the velocity gradient and the influence of the geometry and power consumption of the stirring device in beaker stirring experiments related to the operation of the flocculating basins [91]. They have adopted the Re, the Fr and the power numbers as the dominant dimensionless quantities. The analysis of measurement data yielded a relationship between the Re and the power numbers for vessels and rotors of different geometry. In agreement with earlier experimental results of Camp, they arrived at the interesting conclusion that rotors of different design may produce different flow conditions and patterns, but identical velocity gradients G can be obtained by using rotors having identical projected areas.

The Camp number

Rearranging Eq. (98) and introducing the average computed retention time $t = V/Q$

$$\frac{Q}{V} = \frac{1}{t} = \frac{\sqrt{[D/(\eta V)]}}{Gt} \qquad (99a)$$

or in dimensionless form

$$Gt = \frac{V}{Q}\sqrt{\left(\frac{D}{\eta V}\right)} = \frac{\sqrt{[(DV)/\eta]}}{Q} = \frac{Q_d}{Q} \qquad (99b)$$

where the Camp number is $Ca = Gt$, and Q_d is the discharge defined by the flow that is induced by the energy transmitted.

Considering, for example, a basin of length L and volume $V=LBH$, the work performed in unit time by the friction force—the dissipation of energy—is expressed in terms of the hydraulic radius R of, and mean velocity in, the basin as follows [92]:

$$D = Qg\varrho h_v = Qg\varrho\lambda \frac{l}{4R}\frac{v^2}{2g} = \frac{\lambda LBH}{8R}\varrho v^3. \tag{100}$$

Since $V=LBH$, this is related to unit volume as

$$d = \frac{\lambda}{8R}\varrho v^3. \tag{101}$$

Combining Eqs (98) and (101)

$$G = \sqrt{\left(\frac{\lambda v^3}{8\nu R}\right)} \tag{102}$$

where λ is the coefficient of frictional resistance (and must not be mistaken for the scale factor).

It should be noted that in practice the value of D or d involved in the computation of G is determined on the basis of the energy input or of the pressure head and the differential elevation. The convenient method will vary depending on the design of the structure.

From the above it can be realized that energy dissipation defined as $d=D/V=\eta G^2$, specifically the velocity gradient G (along with the Camp number $Ca=Gt$), play fundamental roles in the description of the flow field. As will be perceived from Eq. (99b), the Ca number can also be defined as the ratio of two characteristic discharges, namely the discharge Q_d of flow induced by energy dissipation and the discharge Q introduced into the structure. Moreover, according to Eq. (99a), in the case of $Ca=$ constant, the hydraulic loading and capacity Q/V is affected not only by V and t, but also by D and η. Experimental evidence seems to imply that for various technological operations the values of G and Gt can be optimized.

Assuming that flocculation processes are similar if the Camp number is invariant ($Ca'=Ca''$)

$$\lambda_G = \lambda_t^{-1} \quad \text{and} \quad \lambda_Q = \lambda_{Q_d}. \tag{103a,b}$$

In the case of $\lambda_\eta = 1$ the scale factor of discharges is

$$\lambda_Q = \lambda_D^{1/2}\lambda^{3/2} = \lambda_G\lambda^3 = \frac{\lambda^3}{\lambda_t}. \tag{104}$$

Further

$$\lambda_G = \lambda_d^{1/2} = \lambda_D^{1/2}\lambda^{-3/2} \tag{105}$$

and

$$\lambda_D = \lambda_G^2\lambda^3. \tag{106}$$

For the case of a rectangular basin of $V=LBH$ volume, Eq. (100) yields with $\lambda_g = \lambda_\varrho = \lambda_\lambda = 1$

$$\lambda_Q = \lambda_D \lambda^{-1} = \lambda_D \lambda_*^{-2}. \tag{107}$$

Using instead of λ_D the scale factor λ_G

$$\lambda_Q = \lambda_G^2 \lambda^2 = \lambda_G^2 \lambda^3 \lambda_v^{-2}. \tag{108}$$

According to Eq. (102)

$$\lambda_G = \lambda_v^{3/2} \lambda^{-1/2}. \tag{109}$$

Substituting λ_G

$$\lambda_Q = \lambda_v^3 \lambda = \lambda_v \lambda^2 \tag{110}$$

whence

$$\lambda_v = \lambda^{1/2} \quad \lambda_Q = \lambda^{5/2}. \tag{111a,b}$$

Viewed from the aspect of hydraulics, Eqs (111a, b) imply the validity of the Froude Law, which also follows directly from Eq. (100).

As a second example consider the case of agitated flocculators, where the energy dissipated per unit volume is

$$d = \frac{2\pi g n M_n}{60V} \tag{112}$$

with n denoting the speed of the stirring device and M_n the effective torque.

Combining Eqs (99) and (112), and introducing the similarity transformation parameters, one obtains for $\lambda_g = \lambda_\eta = \lambda_\varrho = 1$; $Ca' = Ca''$

$$\lambda_Q = \lambda^{3/2} \lambda_n^{1/2} \lambda_M^{1/2}. \tag{113a}$$

The implications of the above relationship are of considerable interest in model and even plant studies. Where the aim is to optimize a particular full-scale equipment ($\lambda = 1$), the discharge Q can be found for different combinations of the operating parameters n and M_n (or vice versa). Another possible form of Eq. (113a) is

$$\lambda_Q \propto \lambda_n^{3/2} \tag{113b}$$

in agreement with the conclusions arrived at by Souček and Sindelar [93] using a different approach.

Modified form of the Camp number

In Section 2.1.4 and in connection with Eq. (99b), mention has already been made of the role played by the dimensionless number $Ca = Gt$. Based on the results of Camp, other investigators have succeeded in demonstrating that the progress of orthokinetic coagulation in time is not described unambiguously by the Ca number alone. From experimental results they have inferred the advisability of using the product $CCa=CGt$. Souček and Sindelar have adopted an even more sophisticated approach. Starting from experimental evidence and theoretical considerations, they have introduced the criterion [93]

$$K_r = \Phi CCa = \Phi CGt \tag{114}$$

where Φ is a quantity representing the quality of the suspension.

The rate of flocculation can be demonstrated to increase with the numerical value of K_r, so that the technological effects can be controlled by the mode of operation. For the case of a suspension of given quality (Φ = constant), the following conclusions of practical interest can be arrived at:

(a) A particular K_r value can be maintained under different combinations of operating parameters C, G and t. In other words, within certain limits any of the above variables can be decreased while increasing the others, without altering appreciably the rate of flocculation.

(b) In the range of orthokinetic coagulation G has been shown to display an optimum value, deflocculation taking place beyond a certain point. This imposes a technological limit on the magnitude of G. It should be noted that it is possible to raise this limit by the addition of certain chemicals (e.g. polyelectrolytes) which increase floc stability. In this way, by increasing the value of G at a given concentration C, the retention time t and consequently the volume V of the structure can be reduced.

(c) In contrast to the intensive agitation needed in perikinetic coagulation, slow stirring is indicated in the orthokinetic coagulation stage by approaching the optimal value of G. The mode of operation is influenced beneficially also by the application of flocculation cascades. The advantage of cascade flocculators stems from the fact that by decreasing successively the gradient G, the size of the flocs is increased before the settling stage and thus the efficiency of settling is improved.

(d) Longer retention times and larger tank volumes are liable to result in poorer economic efficiencies, although lower C and G values lead inevitably to longer retention times. In practice, flocculation structures are commonly dimensioned for $t=10–15$ min retention times.

(e) At higher sludge concentrations C, the steepness of the gradient G can be diminished while maintaining the product CGt constant. Vertical flow, sludge blanket clarifiers represent a practical alternative realization of this approach. The use of some chemical is necessary in this case, too. The concentration C can be increased by sludge recirculation.

(f) The product Gt may become the controlling factor in cases where the role of the concentration C is of secondary importance only. This is the situation for instance with horizontal-flow clarifiers in contrast to the sludge blanket types. In this latter case the optimal value of the product CGt should be adjusted. According to Ives, in actual sludge blanket clarifiers the product CGt assumes values ranging from 60 to 120, a typical normal value being 100. Taking as an average $C = 10^{-3}$ into consideration, the corresponding Ca number is $Ca = 10^5$ [94].

Bearing the above in mind, the similarity criterion of flocculation within the validity range of Eq. (114) becomes ($\lambda_\Phi = 1$)

$$\lambda_C = \lambda_G^{-1} \lambda_t^{-1}. \tag{115a}$$

In the case of $\lambda_C = 1$

$$\lambda_G = \lambda_t^{-1} = \frac{\lambda_Q}{\lambda^3}. \tag{115b}$$

Equation (115b) is substantially the invariance criterion of the Ca number. The practical applicability of modelling on the basis of the Ca number has been demonstrated experimentally by Souček and Sindelar [93]. When the Ca numbers were identical in laboratory and prototype conditions, they observed fairly similar coagulation or flocculation processes in the two, irrespective of the size of the flocculation tank and the intensity of stirring (see Figs 17a, b). This kind of similarity is, however, impossible to realize in the range of deflocculation.

The criterion of deflocculation

The hydraulic characteristics, such as the velocity gradient G, affect the size of the flocs forming, their stability and comminution. The latter phenomenon can also be described with the help of the Re dimensionless criterion. This is shown by Fig. 18, which provides within the validity limits of the corresponding kinetic equations information on deflocculation [93]. The $GRe^{-1/2}$ product is consequently a measure of deflocculation. As long as the product remains below a certain value, the residual iron concentration is constant (see the horizontal section of the plot) indicating that in this case deflocculation is negligible. The pertinent experimental conditions are as follows: alkaline medium, the flocculant is iron chloride, $Ca = 400\,000$. [The Re number is defined as $Re = (v_m d)/\nu$ where d and v_m are the diameter and circumferential velocity, respectively, of the paddle agitator.]

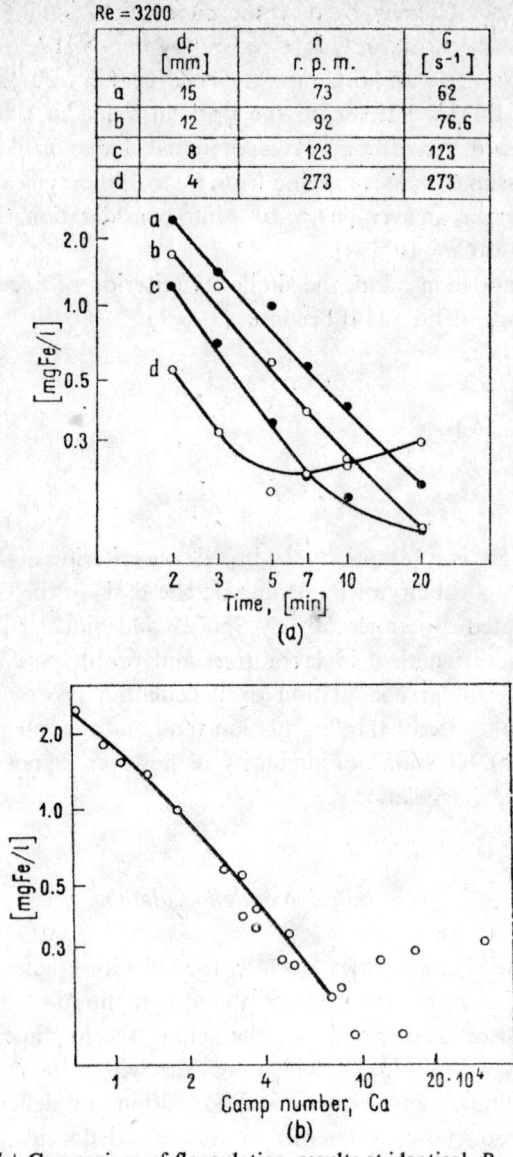

Figs 17(a) and *(b)*. Comparison of flocculation results at identical *Re* numbers (flocculant: iron chloride)

Figure 18 implies the existence of a self-modelling range. In fact, along the horizontal limb of the plot the residual iron concentration is unaffected by the value of $GRe^{-1/2}$.

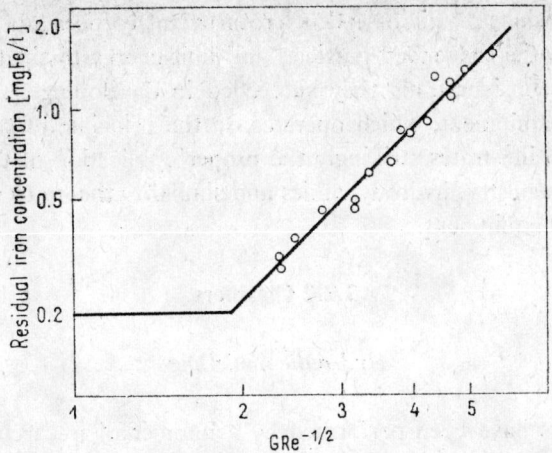

Fig. 18. Deflocculation as a function of the $GRe^{-1/2}$ product

Besides the experimental evidence a theoretical approach can also be adopted to demonstrate the relationship between the Re and Ca numbers [93]

$$Ca = t\sqrt{\left(\frac{E}{Vt\eta}\right)} = \sqrt{\left(\frac{\frac{1}{2}m\bar{v}^2t^2}{Vt\eta}\right)} \propto \sqrt{\left(\frac{\frac{1}{2}\varrho\bar{v}}{\eta} \frac{\bar{v}^2t^2}{l}\right)} \quad (116a)$$

where m is the mass; $\bar{v}$, the mean velocity; and l, the characteristic length. Moreover

$$Ca = \left(\frac{l}{d}\right)^{1/2}\sqrt{\left(\frac{\frac{1}{2}\bar{v}d}{\eta} \frac{\bar{v}^2t}{l^2}\right)} = \left(\frac{l}{d}\right)^{1/2}\left(\frac{1}{2}Re\right)^{1/2}(Ho)^{-1} =$$

$$\text{constant}\left(\frac{l}{d}\right)^{1/2}Re^{1/2}Ho^{-1} \quad (116b)$$

where d is the diameter of the paddle wheel; and Ho, the homochronous number. Finally, the criterion of deflocculation is

$$GRe^{-1/2} = \text{constant}\left(\frac{l}{d}\right)^{1/2}Ho^{-1}. \quad (117)$$

Turbulence and scale-up

Delichatsios and Probstein [95] used a novel approach to the scale-up problem of coagulation, deflocculation and settling. Assuming isotropic turbulence and starting from Kolmogorov's concept, they analysed the typical dimensions of micro- and macroturbulent vortices and the relationship between the suspended particles and unit energy dissipation. Relying on scale-up considerations they succeeded in developing a new type of flocculating equipment, which operates on the principle of tube reactors. This example illustrates strikingly the proper application of the principles of physico-chemistry, hydrodynamics and similarity theory to the design of technological equipment.

2.2.2 Clarifiers

Hydraulic modelling

Model tests have been performed by a number of researchers to study the hydraulic performance of different clarifier types. Some of the tests involving clear water will be mentioned first.

Varrók [96] attempted to improve the flow pattern in the clarifier basin of a thermal power plant in Hungary. The flow pattern was first studied in a laboratory model ($\lambda = 5$) and the silt deposits were checked subsequently in the prototype under normal operating conditions. At normal discharges Re numbers of 2400–2500 were obtained in the cylindrical part of the basin. The flow was classified as turbulent in the majority of cases. Regarding modelling, the aim was to obtain geometrically similar flow patterns. To do this the scale factor λ_Q was varied in the range $\lambda^{5/2} > \lambda_Q > \lambda$. The two limit values correspond to the Froude and Reynolds Laws. Analysis of the results showed that the flow patterns in the model and the prototype were approximately similar geometrically at the scale factors

$$\lambda_Q = \lambda^2, \quad \lambda_v = 1, \quad \lambda_t = \lambda. \tag{118a–c}$$

Therefore the factor λ_Q is closer to the value found on the basis of the Froude Law. Varrók himself considered these results to be of only theoretical rather than practical interest.

Hydraulic studies on a series of settling-clarifier basins have been described by Ivicsics [97]. From the results of tests performed with clear water he arrived at a number of interesting conclusions concerning sound and poor geometrical designs.

It should be emphasized that the results of model tests made with water containing no suspended matter can only be applied with care. Under normal operating conditions the flow pattern, the stability of flow and so on are materially influenced by the presence of the floating sludge blanket, so that the results of clear-water tests are liable to misinterpretation. Nevertheless, the conclusions from such model tests can provide useful guidance in several details of design, such as the comparison of inlet and outflow solutions. The basin design, which clear-water model tests revealed as hydraulically poor, is liable to prove objectionable under operating conditions as well.

Gould's approach

For the hydraulic study of upward-flow clarifiers Gould has developed a scale-up method by which the hydraulic effect of the floating sludge blanket can also be taken into account [98, 99]. It should be emphasized at the outset that his aim was to improve the results of hydraulic model tests made with clear water, rather than to reproduce flocculation and other detail processes in treatment technology. In fact, by creating the floating sludge blanket in the model, a closer similarity of prototype flow conditions is attainable. Following the line of reasoning adopted by Gould, the substance of the method can be summarized as follows.

For the criterion of hydraulic similarity I adopted the invariance of the settling velocity to flow velocity ratio (referred to hereafter as the velocity ratio) and invariance of the Richardson criterion.

The settling velocity of a suspension is defined by the expression suggested by Richardson

$$w^* = w\varepsilon^x \tag{119}$$

where w^* is the settling velocity of the suspension; w, the settling velocity of discrete particles; $\varepsilon = 1-C$, the "porosity" of the suspension, the ratio fluid volume to suspension volume; and x, Richardson's index.

Equation (119) written originally for rigid particles has later been extended for floccular substances as well. (For the case of aluminium hydroxide and fine sand experimental evidence has been provided by Gould.)

As will be recalled, the Ri criterion is

$$Ri = \frac{\Delta\varrho}{\varrho} \frac{gl}{v^2} \propto \frac{g\frac{\partial\varrho}{\partial l}}{\varrho\left(\frac{\partial v}{\partial l}\right)^2} \tag{120}$$

with the quantities involved in the numerator and denominator of the last fraction representing the gradient of stratification and turbulence, respectively. Physically, the Ri number can also be interpreted as the ratio of the (unbalanced) gravitational force $(\Delta\varrho gl^3)$ due to the density differential and the inertial force ($\varrho l^2 v^2$). It follows that the *Ri* number is closely related to the *Fr* number

$$Ri = \frac{\Delta\varrho}{\varrho} Fr^{-1}.$$

Between any two points a and b the value of $\Delta\varrho$ is obtained from the difference in density and concentration as

$$\Delta\varrho = (\varrho_1 - \varrho)(C_a - C_b). \tag{121}$$

It can be demonstrated that if the velocity ratio and the *Ri* criterion are identical in the model and prototype then—assuming the same fluid phase—the following conversion criteria apply [98]

(a) $\quad\quad x' = x''$, i.e. $\lambda_x = 1 \tag{122a}$

the $Re = wd/v$ related to the particle size is invariant ($Re' = Re''$).

(b) $\quad\quad l'^3(\varrho_1 - \varrho)' = l''^3(\varrho_1 - \varrho)''$, so that

$$\lambda_l^3 \lambda_{\Delta\varrho} = 1 \tag{122b}$$

(c) $\quad\quad \dfrac{(\varrho_l - \varrho)'}{v'^3} = \dfrac{(\varrho_l - \varrho)''}{v''^3}$ and $\lambda_{\Delta\varrho} \lambda_v^{-3} = 1. \tag{122c}$

Together with the criteria under (b) and (c) the invariance of the *Re* number related to the whole system also follows. In addition to this, a further similarity criterion can be specified, namely that the propagation velocity u of the change in suspended concentration can be proportionate to the settling and flow velocities. This criterion implies the similarity of concentrations in the model and prototype.

According to Kynch's settling theory

$$u = -\frac{d\Phi}{dC} = w(1-C)^{(x-1)} [C(X+1)-1] \tag{123}$$

where Φ is the flux of the settling matter; and C the volumetric concentration.

The ratios λ_u/λ_w will consequently be equal if $\lambda_C = 1$ and $\lambda_x = 1$, adding that the magnitude of x is influenced primarily by the *Re* number related to the settling particles and by the shape factor.

The scale factors of interest can now be found with the help of the invariance criteria. The criterion given by Eq. (122a) follows from Eq. (123). Consequently ($Re' = Re''$)

$$\lambda_w \lambda_d = 1. \tag{124}$$

In the validity range of the Stokes Law

$$\lambda_w = \lambda_{(\varrho_1-\varrho)}\lambda_d^2. \tag{125}$$

Combining Eqs (124) and (125) the following is obtained

$$\lambda_d^3\lambda_{(\varrho_l-\varrho)} = \lambda_l^3\lambda_{\Delta\varrho} = 1 \tag{122b}$$

as well as

$$\lambda_{(\varrho_l-\varrho)}\lambda_w^{-3} = \lambda_{\Delta\varrho}\lambda_w^{-3} = 1 \quad \lambda_{\Delta\varrho} = \lambda_w^3. \tag{122c}$$

The condition for the *Ri* number to be invariant is thus ($\lambda_\varrho = \lambda_g = 1$)

$$\lambda_{\Delta\varrho}\frac{\lambda_l}{\lambda_v^2} = \lambda_v\lambda_l = 1 \tag{126}$$

implying at the same time—as mentioned before—the invariance of the *Re* number $=(vl)/v$. It is of interest to note that this conclusion has been arrived at by considering the hydraulics of the suspension, rather than the hydraulics of the structure as in the conventional approach. The scale factors of the main geometric and physical quantities are given in Table 3.

Table 3

Scale factors involved in the hydraulic similarity of upward-flow clarifiers

Variable		Scale factor
Designation	Symbol	
Linear dimensions	l	$\lambda_l = l'/l'' = \lambda$
Particle size	d	$\lambda_d = d'/d'' = \lambda$
Velocities	$v;\ w$	$\lambda_v = v'/v'' = w'/w'' = \lambda^{-1}$
Differential density	$\varrho_l - \varrho = \Delta\varrho$	$\lambda_{\Delta\varrho} = \dfrac{(\varrho_l-\varrho)'}{(\varrho_l-\varrho)''} = \lambda^{-3}$
Particle density	ϱ_l	$\varrho_l'' = \varrho + \lambda^3(\varrho_l-\varrho);\ \varrho' = \varrho'' = \varrho$
Time	t	$\lambda_t = t'/t'' = \lambda^2$
Discharge	Q	$\lambda_Q = Q'/Q'' = \lambda_F\lambda_v = \lambda$
Velocity gradient	G	$\lambda_G = G'/G'' = \lambda^{-2}$

Note: The *Ri* and *Re* criteria can be met simultaneously by observing geometric similarity. Settling obeys the Stokes Law.

In the conventional formulation of mechanical similarity, the model law of Reynolds is valid if geometric similarity is maintained. Since the starting basis was the *Ri* number, which is proportionate to the *Fr* number, this result can be obtained by distorting the density ϱ_1 in a convenient manner.

Gould has demonstrated that from the viewpoint of hydraulics the aluminium hydroxide flocs can be successfully replaced with fine sand. The dimensions of the model, or the scale factor λ, are expressed conveniently by the correlation of densities

$$\lambda = \sqrt{\left(\frac{\varrho_l'' - \varrho''}{\varrho_l' - \varrho'}\right)}. \tag{127}$$

For example, if $\varrho_l'' = 2.65$ g/cm³ (sand); $\varrho_l' = 1.002$ g/cm³ (floccular matter); and $\varrho' = \varrho'' = \varrho = 1$, then $\lambda = 9.31$—a truly realistic value. Consequently, in the case of $d' = 1$ mm, $d'' = 0.107$ mm. In a subsequent paper Gould illustrated the application of the theory to a particular problem [100].

It should be added finally that Gould's approach may serve, evidently, also as the starting basis for scaling-up settling tanks.

Technological effect of the floating sludge blanket

The development and effect of the floating sludge blanket can be described in terms of the fundamental flocculation theory. Concerning the water and wastewater treatment implications of this subject, mention must be made of the investigations by Tesařik [101]. Describing some unit processes by dimensionless numbers and expressions, Tesařik has appraised these by dimensional analysis, but made only brief reference to potential applications in scale-up problems. He has concluded that mechanical similarity of flow around settling spherical particles can be described by the Re or Ar criteria, provided that the relationships of Eqs (16a, b) apply [102]. It should also be noted that in the validity range of the Stokes Law the ratio Fr/Re is of decisive importance (see also Section 2.1.3).

Concerning the specific problem of clarifier scale-up, Tesařik has established that in the case of basins having a funnel-type bottom, one essential criterion of mechanical similarity—viewed from the aspect of flocculation—is that the thickness of the sludge blanket be the same in the different systems. As an additional criterion he specified the same $F_1 : F_2 = 0.03$ ratio in the model and prototype, where F_1 is the area of the inlet port and F_2 the horizontal cross-sectional area of the clarifier at the top. This latter criterion applies, obviously, alone to the particular clarifier design that was investigated by Tesařik [103].

Main clarifier dimensions

The considerations above already bear on determining the main clarifier dimensions. Owing to the special importance of the problem in modelling, reference to other research appears to be justified.

The conclusions of Ives [104] may be summarized as follows. The depth of clarifiers and sludge blankets should be envisaged in the model in conformity with the prototype dimensions ($\lambda_h = 1$). Basins having parallel side walls are unsuited to the modelling, of basins with a conical bottom (and vice versa), since the sludge concentrations develop differently in the two systems over time and in space alike. To attain similarity of the sludge distributions and concentrations in conical clarifiers, the side walls should be inclined at the same angle α in the model and prototype ($\lambda_\alpha = 1$). The scale factors $\lambda_h = 1$ (similarity criterion of flocculation) and $\lambda_\alpha = 1$ (similarity criterion of concentration distribution) correspond to the prototype $\lambda = 1$.

Of course, in the case of clarifiers with parallel walls, the dimensions perpendicular to the direction of flow (the horizontal dimensions) can be reduced to the certain limit. For the magnitude of the limit, reference should be made to the work of Pervov [105].

The aim of Pervov's series of experiments was to find the best diameter of the model of a sludge blanket clarifier. He started from the fact that different researchers used a wide variety of sizes (diameters) in their laboratory studies on clarifiers; i.e. Kurgaev: $d = 30$–50 mm, Asanin: $d = 50$ mm, Tshernova: 100×100 mm, Sahov: $d = 100$ mm, Tesařik and Mackrle: $d = 200$ mm. From the laboratory data obtained these authors derived—incorrectly—design criteria for prototype equipment without introducing any scale-up considerations or corrections. To investigate this problem Pervov conducted a series of laboratory experiments, using clarifiers of 3.3 m height and 17, 32, 51, 61, 80 and 200 mm diameter. The main results of this program can be summarized as follows:

(a) The sludge concentration vs basin diameter D'' is described by a hyperbolic relationship

$$C' = C'' + \frac{10^5 k}{D''^2 v} \tag{128}$$

where C' is the concentration of the suspended matter in the prototype clarifier (mg/l); C'' the concentration of suspended matter in the model clarifier; the asymptote of the hyperbola indicating concentration, where the effect of the model diameter D'' becomes negligible, or disappears (mg/l); v, the mean (rising) velocity of flow (m/s); D'', the diameter of the model (mm); and k, a parameter whose magnitude depends on the quality of inflow water and on the flocculants added (e.g. moderately turbid water + aluminium sulphate: $k=0.9-1.2$).

(b) The vertical dimension of the model is controlled by the corresponding dimension of the prototype. In devices of 200 mm diameter the concentration of the suspended medium is not practically influenced by the model diameter.

(c) In the models of the smallest dimensions (17 and 32 mm) no sludge blanket developed, and in some of them channelling, in the others piston flow, was observable. These phenomena are familiar in the technology of fluidized beds.

It should be noted that Eq. (128) can also be used to express the scale factor of concentrations

$$\lambda_c = \frac{C'}{C''} = 1 + \frac{10^5 k}{C'' D''^{1/2} v}. \tag{129}$$

The criterion of $\lambda_C = 1$ is practically

$$\frac{k}{C'' D''^{1/2} v} \approx 0 \tag{130}$$

implying again the necessity of using a sufficiently large diameter D''. Moreover, in adopting the diameter D'' the values of C'' and v must also be taken into consideration. Quite obviously, at $C'' \to 0$ for example, the meaning of Eq. (130) is obscured.

From the above it can be appreciated that the model dimensions to be adopted also depend on the study. In purely hydraulic studies the conventional modelling principles may be adhered to. In contrast, technological processes, such as clarification, should be studied in virtually pilot-scale installations. For scale-up problems associated with such bipurpose or multipurpose studies reference should be made to additional literature on the subject [106].

2.2.3 Activated-carbon adsorption devices

To my knowledge no detailed attention has been given to the scale-up problems related to activated-carbon adsorption devices, even though these play an increasingly important role in water and wastewater treatment technology. The advances achieved in the analysis of transport coefficients, and especially the correlations written in dimensionless form, may nevertheless create the starting basis of defining the scale-up criteria. In this context an example related to the treatment of phenolic wastewater with powdered activated carbon will be described, based on the results of Letterman and his research team [107].

Dimensionless correlations

As the dimensionless expression correlating the film transport coefficient and the physical, hydrodynamic variables, the familiar Gilliland–Sherwood [108] and Frössling [109] equations are widely used

$$Sh = K Re^a Sc^b \quad \text{(for } Sh \gg 2\text{)} \tag{131}$$

and

$$Sh = 2 + K^* Re^c Sc^d \tag{132}$$

where

$$Sh = \frac{K_L d}{D} \quad Re = \frac{dv}{v} \quad Sc = \frac{v}{D} \tag{133a-c}$$

where K_L is the film transport coefficient; d, v, characteristic dimension and velocity; a, b, c, d, experimental constants (exponents); and K, K^*, experimental constants (factors).

Experience has shown the relationship between the Sh, Re and Sc numbers to describe the process fairly adequately. It should be noted that the Re number is often introduced in the following forms

$$Re = \frac{\varepsilon d^4}{v^3} \quad \text{Sherwood and Brian (see Letterman [107, 107a])} \tag{133d}$$

and

$$Re = \frac{\varepsilon^{1/6} d^{2/3}}{v^{1/2}} \quad \text{Calderbank and Moo-Young (see Letterman [107, 107b])} \tag{133e}$$

where ε is the turbulent energy dissipation related to unit fluid volume (based on Kolmogorov's theory of isotropic turbulence) (cm²/s³).

The magnitude of ε can be found experimentally. The energy dissipated in a mixing device, or the unit energy input (which is a quantity proportionate to the velocity gradient G—see Eq. (112) is

$$\varepsilon = \frac{2\pi g n M_n}{60 V \varrho} \tag{134}$$

where n is the speed of the mixer; M_n, the net torque; and V, the fluid volume.

The $\varepsilon = f(n)$ correlation can be determined as a rating curve. From dimensional analysis and also by experimental evidence ε is proportionate to n^3.

Applications to phenolic industrial effluent

Letterman and his co-workers have studied the adsorption effect of activated carbon on the removal of organic material present in effluent under the following experimental conditions: the substance to be removed is phenol at pH=8.3, temperature 20±1 °C, the density of the activated carbon $\varrho_l=0.75$ g/cm³. The diffusion coefficient of phenol $D=0.88\times 10^{-8}$ cm²/s (from the Wilke–Chang equation).

By analysing the experimental results they found the empirical models of both Gilliland–Sherwood and Frössling to ensure the correlation of the required accuracy. In this particular case

$$\frac{K_L d}{D}=0.77\left(\frac{\varepsilon d''}{v^3}\right)^{0.159}\left(\frac{v}{D}\right)^{0.333} \tag{131a}$$

and

$$\frac{K_L d}{D}=2+0.64\left(\frac{\varepsilon d''}{v^3}\right)^{0.197}\left(\frac{v}{D}\right)^{0.333} \tag{132a}$$

where d is the characteristic length, the diameter of the activated carbon particles. An important implication concerning scale-up is that by introducing the term ε, the factor K_L becomes independent of the mixer dimensions.

The validity of these expressions is limited to the case where the size of the activated carbon particles is very small in comparison with the mixer dimensions, i.e. where $\varrho_l/\varrho=0.8-1.25$.

Scale-up criteria

In deriving the similarity criteria, the relationship of Gilliland–Sherwood is adopted as the most convenient starting basis.

Assuming that

$$\lambda_D=\lambda_v=1$$

the scale factor of K_L is found from Eq. (131a) as

$$\lambda_{K_L}=\lambda_\varepsilon^{0.159}\lambda_d^{-0.364}. \tag{135a}$$

Specifying further that K_L should have the same value in the systems of different size ($\lambda_{K_L}=1$)

$$\lambda_\varepsilon=\lambda_d^{2.289}. \tag{135b}$$

Therefore it follows from the criterion $\lambda_\varepsilon=1$ that $\lambda_d=1$ and vice versa.

It should be noted that λ_ε can be calculated using Eq. (134) with operational parameters ($\lambda_g = \lambda_\varrho = 1$, and $\lambda_V = \lambda^3$)

$$\lambda_\varepsilon = \lambda_n \lambda_{M_n} \lambda^{-3} = \lambda_n^3 \lambda^2 \tag{136a}$$

where evidently the dimensions and operating characteristics of the basin and the mixer cannot be neglected any longer. For given basin and mixer dimensions ($\lambda = 1$) the scale factor

$$\lambda_\varepsilon = \lambda_n^3 \tag{136b}$$

pertains to different speeds, which is, of course, consistent with the relationship mentioned in connection with the calibration curve.

2.2.4 Extraction equipment

Phosphorus removal as a potential unit operation in advanced wastewater treatment is quoted as an example, also including the recovery and recycling of the aluminium sulphate (alum). The problems arising were investigated experimentally by Cornwell and Zoltek [110] whose results covered also some domains of modelling. The chemical recovery unit was designed for a 3785 m³/day capacity plant on the basis of laboratory studies.

The process of chemical extraction was generally found to be more efficient in the prototype than in the model. At the same time problems were encountered under operating conditions, which the laboratory experiments failed to predict. Unfortunately, the authors made no reference to the particular scale-up method adopted, mentioning only the works of Ryon *et al.* [111] and Treybal [112] in which the problems of modelling are dealt with more in detail. Treybal suggested, for instance, mixer–settler extraction equipment not exceeding 25 times the size of the model dimensions.

Mention is finally made of the results of Milbury and Stack [113] who offered information on the effects of mixing configurations in laboratory and pilot-scale units used for phosphorus removal at activated-sludge treatment plants. A more detailed explanation about the mutual and positive correlations of the processes taking place at different scales and the scale-up method suggested would permit these statements to be appraised correctly for their practical value.

2.2.5 Elutriation

The biological–chemical methods of phosphorus removal are dealt with in the United States Environment Protection Agency's report prepared by Drnevich [114] on the operation of treatment plants using the activated

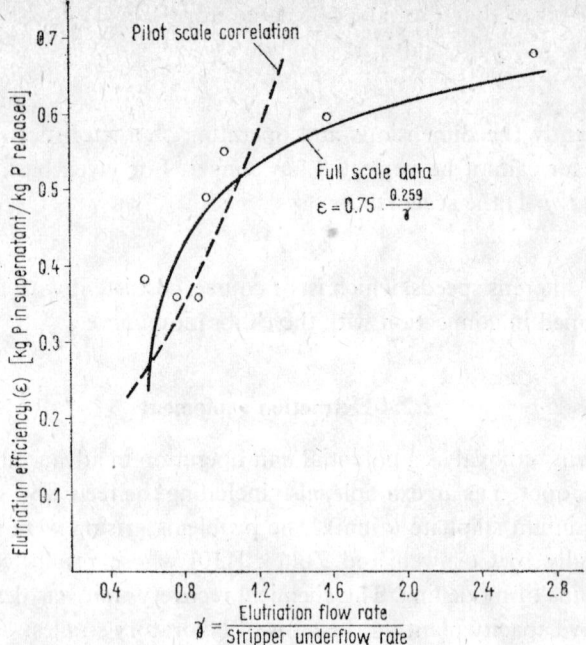

Fig. 19. Elutriation efficiency for the LPE modification

sludge process. The effectiveness of the process referred to as PhoStrip was studied in pilot and full-size units. Here attention is made to the parts of the report in which the processes in the different sizes are compared.

As an illustrative example Fig. 19 shows the variations of elutriation efficiency for the case of the low phosphate elutriation (LPE) system. The data on the pilot and full size units can be seen to agree in a certain range only, beyond which the discrepancy is (probably) due to scale effects.

2.2.6 Disinfection units

Principles of modelling

In connection with model studies on chlorine contact chambers the following model laws are suggested as potentially applicable in the literature on the subject:

(a) Most authors emphasize the applicability of the Froude Law [115], but usually quote no experimental evidence to substantiate their opinions.

(b) Little information is given on the modelling methods based on the Reynolds Law.

(c) Little information is given on the assumption of equal average detention times $\lambda_t = 1$ [116].

(d) One of the most interesting projects in this field has been undertaken by Hart and Gupta [115] who used tracer techniques to determine the flow-through hydrograph; they named as the method the data output similitude method.

The fact that chlorination chambers are as a rule designed for open flow in which gravitational effect is predominant would theoretically support the application of the Froude Law. Identical, e.g. turbulent, flow conditions in the model and prototype are not guaranteed in this way, however. Efforts at complying with this generally recognized requirement may lead to the application of the Reynolds Law. Moreover, where in addition to the hydraulic conditions importance is attached to the reproduction of the technological process, in this particular case disinfection, the observation of the $\lambda_t = 1$ criterion may also be justified, indicating that identical contact times are considered essential in the two systems. This approach is evidently analogous to that followed in modelling other open treatment structures, such as settling basins.

Modelling based on the flow-through characteristics

The idea of using the data of tracer studies, adopted successfully in a number of domains for solving model conversion problems, appeared logical. This attempt is not a novel one, since Vágás [46] in his model studies mentioned earlier on the hydraulics of Dortmund-type settling basins adopted the hydraulic characteristic based on continuous tracer feed, the flow-through hydrograph, for checking experimentally the validity of the model law used.

Hart and Gupta used the flow-through hydrograph, determined after injecting a dye pulse, as the basis of appraisal in their studies. The principle underlying this method is briefly: in similar systems the flow-through hydrographs represented by the dye curves are similar in shape. Moreover, in a dimensionless plot they should be identical. Consequently, the dye curves determined in model tests yield information about the behaviour of the prototype system. The similarity invariants used in model theory are also easily interpreted readily, especially the dimensionless quantities widely used when analysing the properties of the dye curve in hydraulic practice.

These include the ratio, t_{50}/t_i = time, when 50% of the dye is recovered/ideal detention time; t_g/t_i = time corresponding to the centroid of the dye curve/ideal detention time; t_p/t_i = time of peak dye level in the dye curve/ideal detention time; m = dead space fraction; p = plug flow fraction, etc.

The chlorination chambers studied by Hart and Gupta were rectangular in plan and provided with baffles. The model was a geometrically similar replica of the prototype scaled down to $\lambda = 8.43$. The purpose of the measurements in the model and prototype was to determine the model discharge $Q_m = Q''$ at which the hydraulic conditions in the model were similar to those in the prototype, at the particular operating discharge $Q_p = Q'$. To do this the relevant dimensionless quantities were correlated with the discharge Q_m, processing the measurement data by correlation computation. They found the following linear correlations to apply with fairly general validity

$$\begin{aligned} t_{50}/t_i &= aQ_m + b \\ t_g/t_i &= cQ_m + d \\ m &= eQ_m + f \\ p &= gQ_m + h \end{aligned} \qquad (137\text{a--d})$$

Similar expressions could also be written for other dimensionless dependent variables. The empirical constants $a, \ldots, h$ were evaluated by varying the discharge Q_m over a wide range. Conversely, assuming the dimensionless quantities t_{50}/t_i, t_g/t_i, m, p, etc. obtained for the prototype discharge to be identical in the two systems, the corresponding Q_m values were calculated by the empirical expressions, Eqs (137a–d). As originally assumed, the use of the Q_m obtained is a prerequisite for the approximate similitude of flow conditions.

The Q_m values calculated for 12 alternatives with altogether 20 results were tabulated and only three points showed important differences. The Q_m values were determined from a large number of measurements. The results of both graphical and numerical processing revealed that each of the linear correlations of Eqs (137a–d) implied the existence of two ranges as the discharge Q_m was varied. Accordingly, two alternatives were obtained for the $t_{50}/t_i = f(Q_m)$ relationship, one for the lower, one for the higher range of discharge. It is of interest to note that the correlation obtained for the lower range was the closer of the two.

Hart and Gupta arrived at the following conclusions:

(a) The results of model studies are less reliable in the case of models scaled on the basis of one of the dimensionless invariants, such as the Fr number, than in models dimensioned by the method suggested.

(b) The analysis of the experimental data implied the advisability of performing the model tests with discharges lower than the $Q_m = Q''$ discharge calculated using the Froude Law, since this is likely to result in a more reliable and accurate model—prototype correlation.

(c) The characteristic quantities commonly used in tracer studies differ from each other in reliability. For this reason the most representative or suitable quantities must be selected for each model study by considering the circumstances in advance. As revealed by these tests, the dimensionless quantities t_g/t_i, t_{50}/t_i, p, m, as well as the standard detention efficiency (SDE) and the fraction dead volume (FDV) proved most effective in modelling this particular prototype.

The work of Hart and Gupta has been discussed at length because of the interesting observations they presented.

Further processing and analysis is possible in an attempt to decide which of the dimensionless invariants characterizing partial dynamic similarity is most suited for use in modelling.

Without embarking upon the details of the analysis, only the principal steps will be outlined here. The normal sewage flow in this particular case is known to be $Q_p = Q' = 8.6$ Mgal/day, approximately 6000 gal/min. It has been established that the model discharge $Q_m = Q'' = 31$ gal/min is the average for the various alternatives mentioned, of which the three giving extreme values have been neglected.

The problem consisted of finding the approximate value of the exponent α_Q involved in the scaling equation

$$\lambda_Q = \frac{Q'}{Q''} = \lambda^{\alpha_Q} \tag{138}$$

with the known values of Q', Q'' and $\lambda = 8.43$. The result is $\alpha_Q = 2.47$, which is in surprisingly good agreement with the theoretical exponent 2.5 corresponding to the Froude Law.

It can be concluded that the approximate validity of the Froude Law is substantiated by the results of the tracer study. Though Hart and Gupta have failed to recognize this implication, the result obtained does not conflict with their statements. The recommendation of a model discharge which is smaller than that corresponding to the Froude Law is probably due to the fact that they ascribed—in view of the correlation coefficients obtained —greater significance to the lower of the two discharge ranges.

The simplified nature of the method adopted for analysis is recognized and could be made more exact by including the specific data available to the authors. Nevertheless, the agreement with the Froude Law seems convincing. In this respect reference is again made to the tracer studies of Vágás on the modelling of Dortmund-type settling basins, which gave a similar value $\alpha_Q = 2.476$ (see the subsection on the modelling of vertical flow settling basins, in Section 2.1.3).

In conclusion it is deemed necessary to call attention to the great number of publications in the professional literature, which contain interesting information concerning hydraulic model studies on chlorination units, with brief references to model-prototype conversion problems. A typical example is the studies of Louie and Fohrman [117] which aimed to find the efficient design of chlorine mixing and contact chambers in a model built to the scale factor $\lambda = 10$ using the distributions of flow velocity and detention time.

2.3 BIOLOGICAL TREATMENT

2.3.1 Trickling filters

The scale-up problems associated with trickling filters can be analysed by observing principles similar to those developed for modelling packed columns. A few potential applications to practical problems will be outlined, subsequently drawing on the literature on the subject.

Simulation of trickling filters

Mathematical and physical models have been devised by several authors to reproduce the processes taking place in trickling filters. The phenomena on a suitably inclined surface have often been adopted as the basis of simulation. The efforts of Maier *et al.* [118], Swilley and Atkinson [119] and Quirk [120] are mentioned as examples. Quirk suggested a conversion from the inclined plane to the packed column based on the surface loading, rate T_s

$$T_s = \frac{Q}{B/a} \tag{139}$$

where T_s is the surface loading rate (m/h); Q, the discharge over the inclined plane (m³/h); B, the width of the inclined plane (m); and a, the unit surface area of the packing medium (m²/m³).

By introducing the surface loading rate or the average detention time into the kinetic equations of organic removal, Quirk has succeeded in writing the expressions applying to the prototype, the validity of which has been verified experimentally.

Description by dimensionless relationships

No uniform form for the dimensionless numbers of hydraulic (dynamic) meaning used in investigations on trickling filters has yet been recognized. In the case of the simulation model mentioned in the previous subsection, where the flow over a plane inclined at the angle α to the horizontal is considered, the relevant Fr and Re numbers can be written into the following forms [121]; for example

$$Fr = \frac{v_{ave}^2}{g\delta} \qquad Re = \frac{4Q}{\nu B} \qquad (140a,b)$$

where the average flow velocity v_{ave}

$$v_{ave} = \frac{g\delta^2 \sin \alpha}{3\nu} \qquad (141)$$

and δ is the thickness of the fluid film and $\nu = \eta/\varrho$, the kinematic viscosity of the fluid.

Swilley and Atkinson [119] have combined the hydraulic and physico-chemical characteristics of trickling filters in their dimensionless relationships. Their investigations have shown the following dimensionless quantities to play major roles: the Re number, the Sc number, the Packing number (Pa), peculiar to the packed column and the Reaction number (Ra), peculiar to the conditions under which the reaction takes place. By an experimental approach they have determined the relationship between the main dimensionless quantities – mathematical model of the trickling filter – for different operating conditions, such as laminar and turbulent flow.

The work of Sinkoff *et al.* [122] is a similar example confined to the purely hydraulic aspects in which a dimensionless expression has been written, correlating the average retention time, the hydraulic loading rate, the specific surface and the depth of packing as the main operational parameters. The method used was dimensional analysis [122].

In describing wastewater treatment by trickling filters Gerber [123] introduced dimensionless quantities and correlations. The representative expression, the dimensionless formula

$$\frac{C}{C_0} = f\left(\frac{q}{Hk_1}\right) \qquad (142)$$

is given here where the physical quantities are C_0, C, the BOD$_5$ concentrations of the water before and after passing the trickling filter, respectively (mg/l); q, the daily hydraulic loading rate (m³/m²day); H, the depth of the trickling filter (m); and k_1, first-order reaction rate constant (day^{-1}).

Yakovlev and Galanin [124] have adopted a similar approach, introducing the value 0.4 for the exponent of q involved in Eq. (142) to achieve better agreement with actual experience. It should be noted at the same time that the ratio $q^{0.4}/(Hk_1)$ is no longer a dimensionless number.

The analysis of Tuček et al. [126] on trickling filters by dimensionless numbers is probably the most detailed one so far presented. Starting from the results of Gerber, they attributed fundamental importance to two dimensionless quantities derived by dimensional analysis (with results substantially comparable to those of Gerber):
— for the case without recirculation

$$\pi_1 = \frac{C_0 - C_2}{C_0} = \frac{\Delta C}{C_0} \qquad \pi_2 = k_1 \frac{H}{q} \qquad (143a,b)$$

— for the case with recirculation

$$\pi_{1R} = \frac{C_{0R} - C_2}{C_{0R}} = \frac{\Delta C}{C_{0R}} \qquad \pi_{2R} = k_1 \frac{H}{q_R} \qquad (144a,b)$$

where C_0, C_2 are the BOD$_5$ concentrations in the inflow to the trickling filter and in the effluent from the secondary settling tank, respectively (mg/l). The subscript R refers to the quantities changed as a consequence of recirculation.

By processing a large number of observations (using those of Keefer and Meisel, and Tomlinson and Hall), Tuček and his co-workers [126] found the following expressions to apply

$$\pi_1 = \frac{\pi_2}{a + b\pi_2} \qquad \pi_1 = \frac{\pi_2}{1.07(0.089 + \pi_2) - 0.09}. \qquad (145a,b)$$

The constants a and b determined experimentally are $a = 0.089b - 0.09$; $b = 1.07$.

It is of interest to note that the experimental constants have the same value in the devices operated with and without recirculation. Tuček et al. gave an example of dimensioning to show the application of the above going expressions. Unfortunately they omitted to specify the validity ranges of the expressions suggested.

The role of turbulence

Recently Hartmann [127] studied, under laboratory conditions, the impact of turbulence (in terms of the *Re* number) on the activity of the biological film. The film was developed on the internal surface of a 1.0 m glass tube having a diameter of 1.0 cm. The experimental conditions were as follows: the mean velocity of flow in the tube $v_m = 14-107$ cm/s, temperature $T = 22-23$ °C, the kinematic viscosity of the medium: 1.3×10^{-6} m²/s. Assuming the reaction kinetic model of Michaelis–Menten to apply, Hartmann determined the highest value of the reaction rate r_{max} and the Michaelis constant K_m as kinetic parameters for different flow velocities. He found the value of the *Re* number—the extent of turbulence—to affect materially the rate of reaction in the aerobic biological system. This influence is due primarily to the mass transport of nutrients to the bacterial cells. Nutrient transport may be a limiting factor in the chain of reactions. The rate of oxygen transport is also modified by turbulence. The kinetic constants depend implicitly on the prevailing flow conditions and this fact should be allowed for in scale-up considerations.

The flow conditions in the biological film on the packing of trickling filters can be described well—as will be recalled—by dimensionless correlations, which may also have scale-up implications. A specific example is the study published by Williamson and McCarthy in 1976, in which the depth of the stagnant liquid layer as a function of the Reynolds number was examined, and thus the effect of turbulence [128]. The rate of oxygen uptake, the distribution of substrate concentration and the magnitude of other quantities affecting the technological process are controlled by this depth. Evidently, in these investigations a *Re* number defined properly (the *Re* number applying to packed columns) must be introduced.

Scaling-up in trickling filter technology

The problems related to the modelling of trickling filters have been dealt with cursorily by Eckenfelder and Cardenas [129]. They started from the assumption that in scaling up laboratory models of biological systems to prototype size, it is essential to ensure identical biological and hydraulic conditions. They therefore adopted the so-called "trivial" approach to modelling. However, this principle cannot be adhered to unless certain conditions are observed which the authors failed to clarify sufficiently. Eckenfelder and Cardenas have, nevertheless, appraised the model–prototype aspects of the kinetic relationships and paved the way for Quirk's investigations mentioned before.

As long as specific evidence founded on actual experiments is not available in this field, scale-up problems associated with trickling filters can be approached on the basis of the following considerations:

(a) The familiar scale-up and modelling principles applied in chemical process engineering are assumed to retain their validity.

(b) Pilot-scale trickling filters can be designed and operated in accordance with the conclusions arrived at in the review of deep bed filters (see Section 2.1.4: Experience gained from scaling-up experiments, (a)–(d), [76]).

(c) The parameters involved in the kinetic reaction processes can be transferred from one system to the other with the help of the appropriate basic equations [120, 129].

These conclusions are largely consistent with those of Jones [130], who stated that model or pilot-scale studies on trickling filters are possible using devices of the same height H as the prototype ($\lambda_H = 1$), but having a smaller diameter. The same height is conducive to the development of typical zones—in the biological sense—resembling those within the trickling filter. The ratio D/d of the diameter D of the pilot unit to the average particle size d of the packing medium should be higher than 8 : 1 in order to avoid any wall effect. The particle size of the packing medium should not be reduced for the model study ($\lambda_d = 1$). Ignoring this condition may be justified only in experiments serving special purposes, such as hydraulic studies on alternative designs. Finally, the actual prototype surface and volume-loading rates $T_s = Q/F$ and $T_v = Q/V$ respectively, should be considered in both the model and pilot equipment ($\lambda_{T_s} = \lambda_{T_v} = 1$). Consequently in the case of $\lambda_H = 1$

$$\lambda_Q = \lambda_F = \lambda_V. \tag{146}$$

2.3.2 Aeration tanks

Hydraulic similarity

Aeration tanks and basins are the principal structures at activated-sludge biological treatment plants. Two main types of design may be distinguished, one involving some kind of diffused air system, the other involving surface aeration by means of horizontal or vertical-shaft rotors.

For the modelling of conditions in aeration tanks of the INKA type with compressed-air injection, a solution was suggested by Fischerström [131], although without experimental verification. Froude's model law served as the basis of scaling up, without taking the rising velocity w of the bubbles into consideration. Consequently the scale factor of the air discharge introduced can be determined with crude approximation only on the basis of $\lambda_Q = \lambda^{5/2}$ corresponding to the Froude Law. Fischerström considered models scaled down more than 1 : 2 ($\lambda = 2$) to be inadvisable for experimental purposes. For economic appraisal, on the other hand, he suggested prototype-scale systems ($\lambda = 1$) alone.

Robertson [132] adopted one of the principles developed for mixers to reproduce the hydraulic conditions in Simcar-type, vertical-shaft aerator systems. He assumed the power demand N/V related to unit fluid volume to be constant in geometrically similar ($V \propto d^3$) systems of different size

$$\frac{N'}{V'} = \frac{N''}{V''} = \text{constant}_1 \qquad \frac{N}{d^3} = \text{constant}_2 \qquad (147\text{a,b})$$

where N is the power demand; V, the fluid volume mixed; and d, the diameter of the aerator rotor (mixer).

From Eqs (147a, b) it follows that in the case of turbulent flow ($N = $ constant $n^3 d^5 \varrho$)

$$n^{1.5} d = \text{constant}. \qquad (148)$$

Changing over to the similarity transformation parameters, the condition equations of scaling up are

$$\lambda_N = \lambda_V = \lambda^3 \qquad \lambda_n = \lambda^{-2/3}. \qquad (149\text{a,b})$$

The scale factors of velocities and discharges are

$$\lambda_v = \lambda^{1/3} \qquad \lambda_Q = \lambda^{7/3} = \lambda^{2.33}. \qquad (149\text{c,d})$$

It is of interest to note that the exponent 2.33 is in fair agreement with 2.5 resulting from the Froude Law.

Several authors have adopted the invariance of the Fr number as the condition of similarity in modelling the hydraulic conditions in vertical-shaft aeration rotors. The validity of this approach has been verified experimentally by Knop and Kalbskopf [133, 134] for a particular case. Figure 20a is reproduced from their paper. The diagram represents the relationship between the bottom velocity of flow v and the volume V of the aeration basin at the unit power uptake of 20 W/m³ (R_r is the ratio of net basin volume to the wetted surface). The curve applies to a square basin with the

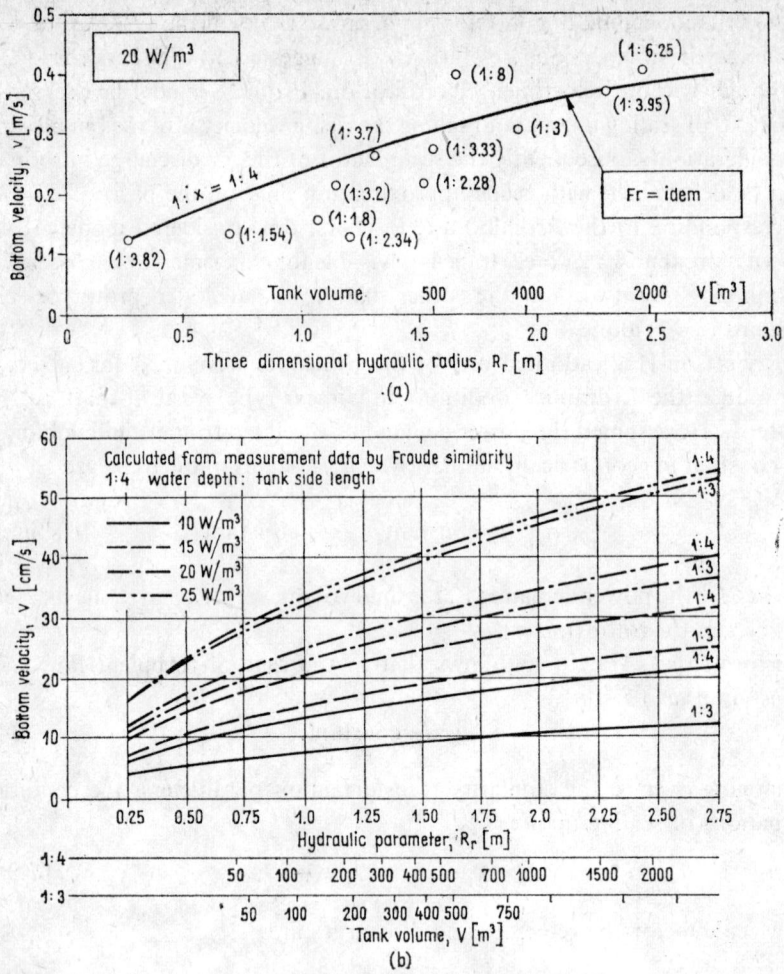

Figs 20(a) and *(b)*. Bottom velocities in vertical-shaft aeration tanks vs net volume and R_r; in tanks with vertical-shaft rotors vs the hydraulic parameter and unit power consumption

ratio $1:x = 1:4 =$ water depth to basin width. The points of the curve correspond in the case of geometrical similarity to the Froude model law. The experimental points that lie close to this ratio also fit the curve, demonstrating the validity of the Froude Law (up to a unit value of 30 W/m³). The effect of the depth : width ratio in the basin is shown by the data differing substantially from these. In the case of relatively shallower basins the experimental data are situated above the theoretical curve, whilst below it in deeper basins. In this sense the comparison is between geometrically distorted systems. Starting from the investigations of Knop and Kalbskopf and processing the experimental results of several authors, the family of curves in Fig. 20b was constructed by Kemény who extended the range of experiments [135].

For similarity of flow patterns the Froude Law was also considered acceptable by Kalinske *et al.* [136], emphasizing that the reoxygenation, mass transfer, etc. processes will not necessarily become similar, however.

Zeper and De Man [137] have used a model approach in developing the Carrousel-type aeration system. Besides geometric similarity, they emphasized the importance of obtaining similar head losses. As an index of hydraulic resistance they introduced the velocity coefficient C involved in the Chézy formula, presenting a method for its experimental determination. They found the resistances to be similar at the model scale $\lambda=20$. At the same time they also assumed the validity of the Froude Law. Zeper and De Man succeeded in correlating the rotor speed to the mean velocity of flow in both the model and prototype. At velocities higher than 30 cm/s the curves obtained in the model differed from those in the prototype, for which they suggested the approximate character of the Froude Law and the scale effect as potential explanations. The latter may be especially important in view of the rather high scale factor of $\lambda=20$. This prompted them to construct a 1 : 10 scale model for subsequent experiments.

Lewandowski *et al.* [138] reported a novel experimental approach to reproducing the hydraulic conditions in vertical-shaft aerators. Starting from dimensional considerations they arrived at the conclusion that the hydraulics in the (Simcar-type) aerator examined are described by the following three dimensionless quantities:

$$Fr^{1/2}=\frac{v}{\sqrt{(gl)}} \quad A=\frac{n^2l}{g} \quad B=\frac{\varrho^2gl^3}{\eta^2}. \qquad (150\text{a–c})$$

It should be noted that the A number is nothing other than the Fr number expressed in terms of the rotor parameters, while the B number is the familiar Galilei number ($B=Ga=Re^2/Fr$).

In the Fr and B numbers the basin size, whilst in the A number the rotor diameter, should be introduced as the characteristic length. Consequently, the Fr and A numbers can be maintained invariant even if the rotor size is distorted (geometric similarity is violated).

The tests made in three models of different size (denoted I, II and III) with five rotor models had the object of determining the correlations between the relevant quantities. The scale ratios of dimensions were $\lambda=2$ for the

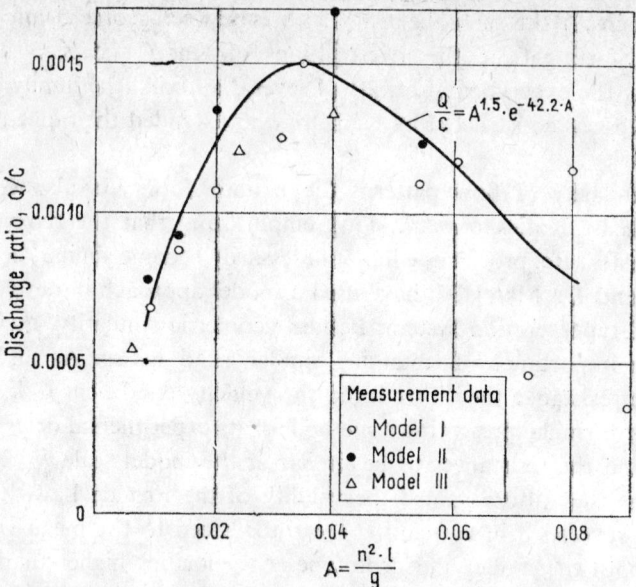

Fig. 21. The $Q/C=f(A)$ relation for three models of different size

models I and II, $\lambda=2$ for the models II and III and $\lambda=4$ for the models I and III. The correlation applying to each of the three models was

$$Q = CA^{1.5} e^{-42.2A} \qquad (151)$$

where Q is the pumping capacity of the rotor, the discharge passing a representative horizontal plane (wetted surface) of the basin (m³/s).

As can be seen from Fig. 21, the curve $Q/C=f(A)$ assumes a peak at $A=0.036$. The factor C depends on the B number and the unit of measurement is m³/s. Equation (151) substantially represents the correlation $\sqrt{Fr} = f(A,B)$. The simultaneous invariance of the Fr and A numbers leads to the expression $\lambda_Q = \lambda^{2.5}$. Moreover, according to Eq. (151) the Q/C ratios in the different models must also be equal. Thus

$$\lambda_Q = \lambda_C = \lambda^{\alpha_Q}. \qquad (152)$$

Calculating the exponent from the data published in their paper, Lewandowski *et al.* have obtained α_Q values higher than 3, which differs significantly from the value 2.5 corresponding to the Froude Law. The values of C for the three models were $C_I = 99.9$, $C_{II} = 11.0$ and $C_{III} = 0.892$ m³/s, attributable, according to the authors, because only the criteria resulting from the Fr and A numbers, without observing the invariance of the B number, have been included. In my opinion an additional explanation can be offered. If one accepts the validity of Eq. (151) for each of the three models and the A number is actually invariant, then the ratio Q/C should also be invariant. Any departure from this can be traced back, at least in part, to the inaccuracy involved in the calculation of both Q and C. It can be readily appreciated that the experimental determination of Q may be afflicted with significant inaccuracies. As a final conclusion Lewandowski et al. suggested the acceptability of the modelling method based on the Froude Law.

Harremoës [139] has also adopted dimensional analysis in his model studies on the hydraulic and mass transfer processes in aeration tanks. He illustrated his theory in connection with the analysis of aeration tanks of rectangular cross section with the air blown in at the bottom. Use was made of the analogy with the theory of the buoyant jet. Starting from there he adopted the quantity $(gQ_0)^{1/3}$ as the velocity characteristic of the system (where g is gravitational acceleration and Q_0 the air supply at the bottom per unit length of tank). Harremoës succeeded in demonstrating (model size 0.4, prototype size 4.0 m) that the circulation created by the air introduced, the surface and bottom velocities, can be correlated with the foregoing characteristic velocity. The main conclusions are:

(a) By dimensional analysis he obtained the following characteristic dimensionless numbers for describing the velocities in the aeration tank
— simplex numbers

$$U=\frac{u}{(gQ_0)^{1/3}} \qquad L=\frac{h}{b} \qquad (153a,b)$$

— complex numbers

$$Fr_Q=\frac{(gQ_0)^{1/3}}{(gh)^{1/2}} \qquad Re_Q=\frac{(gQ_0)^{1/3}h}{v} \qquad (153c,d)$$

where u is the velocity of flow; h, the depth of water in the tank; and b, the tank width.

(b) The characteristic dimensionless correlation would be consequently

$$U=\text{constant } L^a Fr_Q^b Re_Q^c \qquad (154)$$

where the constant and the exponents a, b and c are to be determined by experiments.

(c) By processing the data obtained in the model and prototype he found the effect of the Fr_Q and Re_Q numbers to be negligible in this particular case (i.e. $b \approx c \approx 0$). This statement is, however, unacceptable without further explanation. It is evidently not intended to mean that the Fr and Re numbers are negligible in describing the flow conditions in the tank. Neglect of the Fr number would not be permissible especially in modelling, since this would contradict the repeatedly substantiated findings of several researchers. The only conceivable explanation is that in calculating the dimensionless simplex U as a dependent variable, the role of the Fr_Q and Re_Q numbers is small enough to be neglected.

(d) The dimensionless correlations obtained graphically are:
concerning the circulatory flow in the tank

$$\frac{q/h}{(gQ_0)^{1/3}} = 0.95 \left(1+\frac{h}{b}\right)^{-1.27} \tag{155a}$$

where q is the discharge of the circulated flow (determined from the velocity distribution)
concerning the surface velocities u_0

$$\frac{u_0}{(gQ_0)^{1/3}} = 3.65 \left(1+\frac{h}{b}\right)^{-0.86} \tag{155b}$$

and finally, concerning the bottom velocities u_b

$$\frac{u_b}{(gQ_0)^{1/3}} = 2.5 \left(1+\frac{h}{b}\right)^{-0.86}. \tag{155c}$$

(e) The correlations obtained can be interpreted for, and applied to, aerated grit chambers as well. According to Harremoës, for instance, if $u_b = 0.3$ m/s, $h/b = 1$, $(gQ_0)^{1/3} = 0.25$ m/s, then $Q_0 = 1.6 \times 10^{-3}$ m³/s m $\approx$ ≈ 6 m³/h m.

(f) The model and prototype data are compared in Figs 22a–c showing the reasonable agreement between the two. Both are situated along the fitting line, although the scatter of the prototype data is wider.

As a last example the paper by Maise [140] can be considered. He identified the Re, Fr and We numbers as the bases of the condition equations of dynamic similarity. The familiar fact that the three dimensionless quantities above cannot be made invariant at the same time was emphasized. For this reason he adopted a correlation having the form of a power product of the relevant dimensionless quantities as the starting basis. Founding on experiences gained with industrial mixers he considered a combination of the power number (the Eulerian number), the Re and the Fr numbers to be significant in scaling up the hydraulic conditions in surface aeration systems

$$\frac{N}{\varrho n^3 d^5} = K \left(\frac{nd^2}{v}\right)^a \left(\frac{n^2 d}{g}\right)^b. \tag{156}$$

The constants K, a and b are evaluated from experimental results. Unfortunately, no measurement results were published by the author. Equation (156) is equivalent with Eq. (170) to be considered later.

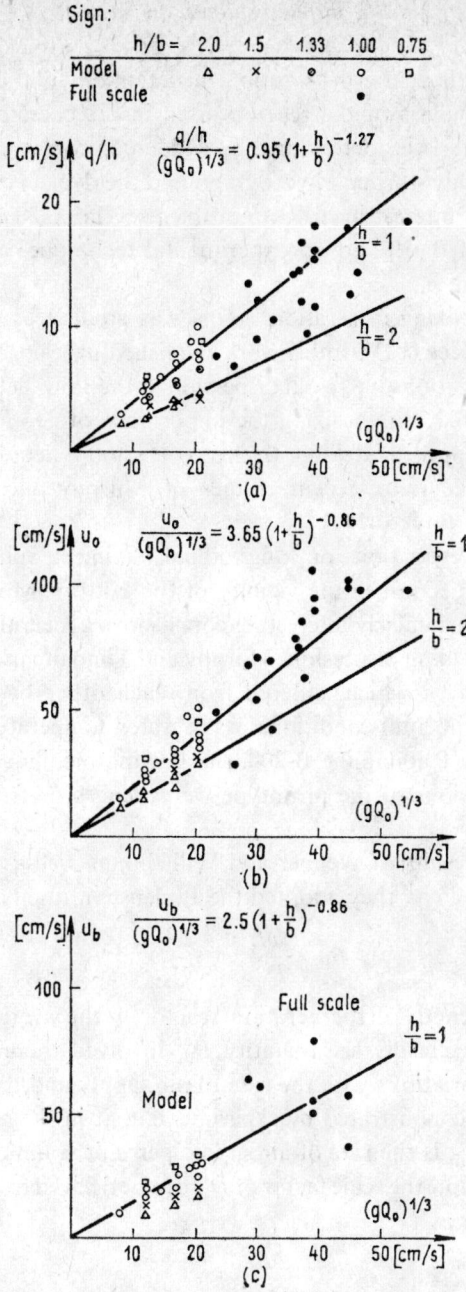

Figs 22(a)–(c). Measured velocities as a function of characteristic velocity for both model and full-scale tanks

Flow-through studies, dispersion

As will be recalled, the distribution of detention time is also a suitable means for the analysis of the reactors used in the chemical industry. The hydraulic characteristics determined by sudden or continuous-dye injection —known as the flow-through hydrograph and the dye curve—may yield information also of interest in scale-up problems. The axial dispersion coefficient D_{ax} can also be found by experimental techniques as a special index number of mixing.

Longitudinal mixing in aeration basins was studied as early as 1944 by Thomas and McKee [141]. In his work published in 1966, McLean [142] reported on investigations into axial dispersion in the flow field of laboratory-scale aeration tanks, emphasizing the importance of geometric design also from the viewpoint of modelling. In this review more detailed attention will be devoted to the more recent studies of Murphy and Timpany [143] and Murphy and Boyko [144].

From their investigations on longitudinal mixing in spiral flow aeration basins with air injection in the vicinity of the bottom, Murphy and Boyko have arrived at particularly interesting conclusions concerning the condition equations of modelling dispersion. Murphy and Timpany used two geometrically similar devices which differed from each other by the scale factor $\lambda = 13$. The experimental conditions were: water temperature $T = 19-20°C$, the dye used was Rhodamine B-200, the flowing medium was wastewater. The main dimensions of the prototype were: length 66 ft, depth 15 ft and width 30 ft.

Based on the results of Wehner and Wilhelm, as well as Levenspiel [297] on reactor technology, they adopted the Bodenstein number

$$Bo = \frac{vL}{D_{ax}} = \frac{L^2}{D_{ax} t} \tag{157}$$

where L is the length of the aeration reactor, v the average flow-through velocity $= L/t$, specifically the quantity D_{ax} involved therein as significant, correlating the variations with the rate of air supply and the average detention time. They demonstrated by experiments that in the case of $\lambda_t = 1$ and $\lambda_{q_{air}} = 1$ (where q_{air} is the rate of air supply per unit volume, thus m³ air/m³ wastewater per hour) the scale factor of D_{ax} can be derived from the invariance of the Bo number as

$$\lambda_{D_{ax}} = \lambda^2. \tag{158}$$

For comparison of the characteristics related to the model and prototype, Figs 23 and 24 are given (after Murphy and Timpany).

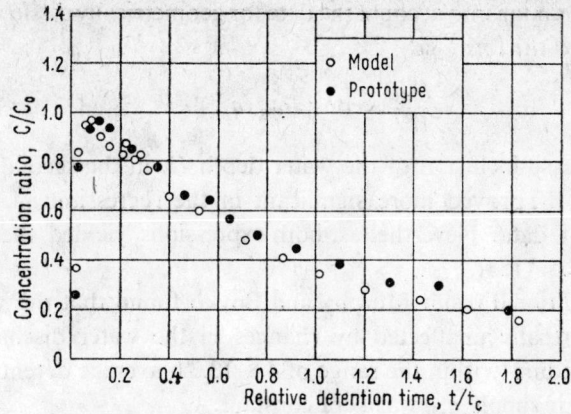

Fig. 23. Model and prototype compared on the basis of the flow-through hydrograph

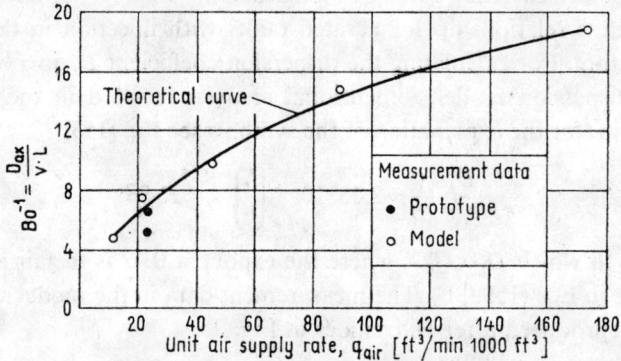

Fig. 24. The Bo^{-1} number vs the unit air supply rate

The distribution of detention times determined in the geometrically similar model and prototype, the flow-through hydrograph, is shown as a dimensionless plot in Fig. 23 for the case where $\lambda_t = \lambda_{q_{air}} = 1$. On the other hand, the q_{air} vs Bo^{-1} relationship is given in Fig. 24. As can be seen any increase of q_{air} entails an increase in D_{ax}, i.e. the intensity of dispersion (mixing). At a particular air discharge q_{air} in two experimental prototype designs, the values 5.2 and 6.4 were obtained for the dimensionless quantity $D_{ax}/(vL)$, so that the difference related to the model value of 7.2 is only slight.

Further evidence for these conclusions was provided by Murphy and Boyko [144]. By analysing a number of experimental designs and measurement data they derived for geometrically similar systems the formula

$$\frac{D_{ax}}{H^2} = \text{constant}_1 \, (q_{air})^{0.461} \qquad (159a)$$

whereas also taking into account the data for geometrically dissimilar systems, they obtained the formula

$$\frac{D_{ax}}{W^2} = \text{constant}_2 \, (q_{air})^{0.346}. \tag{159b}$$

In the first case inclusion of the water depth H, in the second that of the basin width W, proved more significant in the regression analysis of the measurement data. Nevertheless, both expressions yielded the conversion formula of Eq. (158).

As an additional result Murphy and Boyko found that the value of D_{ax} remained virtually unaffected by changes in the water discharge, or the hydraulic loading (within the range of 4.8–13.3 h average detention time) as long as the air supply q_{air} was kept constant.

Harremoës [139], drawing on the experimental data of Murphy and co-workers, analysed the role of longitudinal mixing, specifically the dimensionless relationship for aerated tanks with injection in the vicinity of the bottom. For calculating the dispersion coefficient D_L involved in the familiar dispersion model, dimensional analysis resulted in the empirical relationship (for the explanation of the symbols see Eq. [153])

$$\frac{D_L}{(gQ_0)^{1/3}b} = 2.4 \times 10^{-3} \left(\frac{h}{b}\right)^{-0.68} Re_Q^{0.26} \tag{160}$$

according to which $D_L \propto Q_0^{0.42}$ where the exponent 0.42 is in fair agreement with those in Eqs (159a, b). The measurement data in the model and prototype are reproduced after Harremoës in Fig. 25.

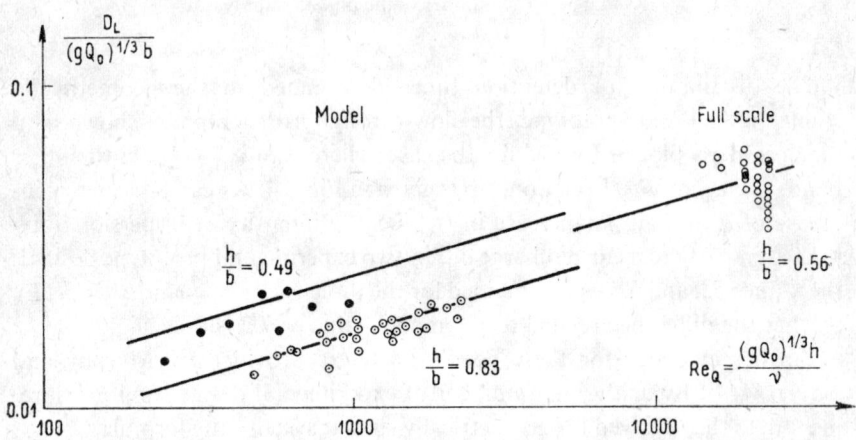

Fig. 25. Longitudinal dispersion in aeration tanks as a function of geometry and Re number for the rotational flow

The inverse of the dependent variable in Eq. (159) may be regarded a modified Bodenstein number. Starting from here I shall examine—by applying the concept of the invariant function introduced earlier—the relationship between the similarity transformation parameters which serves as the basis of modelling. However, it should be remembered that the quantity Q_0 relates to unit tank length, while q_{air} to unit tank volume ($\lambda_{Q_0} = \lambda_{q_{air}} \lambda^2$). Also assuming as in the case of Eq. (158) that $\lambda_{q_{air}} = 1$ and $\lambda_b = \lambda_h = \lambda$, with $\lambda_v = 1$, then

$$\lambda_{D_L} = \lambda_{Q_0}^{0.42} \lambda^{1.26} \tag{161a}$$

or changing over from Q_0 to the variable q_{air} ($\lambda_{q_{air}} = 1$)

$$\lambda_{D_L} = \lambda^{2.1}. \tag{161b}$$

This, although derived by a totally different approach, is in surprisingly good agreement with the transformation formula given as Eq. (158).

The effect of turbulence

The variation in the activity of activated sludge flocs with the intensity of turbulence, i.e. the Re number, was studied under laboratory conditions by Hartmann and Laubenberger [145] who performed experiments by passing a sludge suspension through a tube reactor. The mean velocity of flow varied from 5 to 100 cm/min at temperatures of 15, 20, 25 and 30°C. The rate of biological activity was measured in terms of oxygen consumption expressed as γO_2/mg N per minute (where N is the organic nitrogen contained in the flocs). Figures 26a, b show the results from this analysis. The main conclusions are summarized as follows:

(a) Turbulence acts in two ways (and accordingly the curves in the figure can be divided into two typical sections), on the one hand by accelerating the disintegration of flocs, on the other by promoting the exchange of boundary surfaces. Consequently, oxygen consumption increases together with the Re number.

(b) Floc disintegration may start in the laminar range. At Re numbers higher than about 5000 the physical properties of the flocs are not affected to any significant extent by the type of flow (in this respect a self-modelling range may be inferred to exist).

(c) Exchange reactions between the wastewater and the boundary surface of bacterial cells takes place mainly in the range $Re = 5000$–9000.

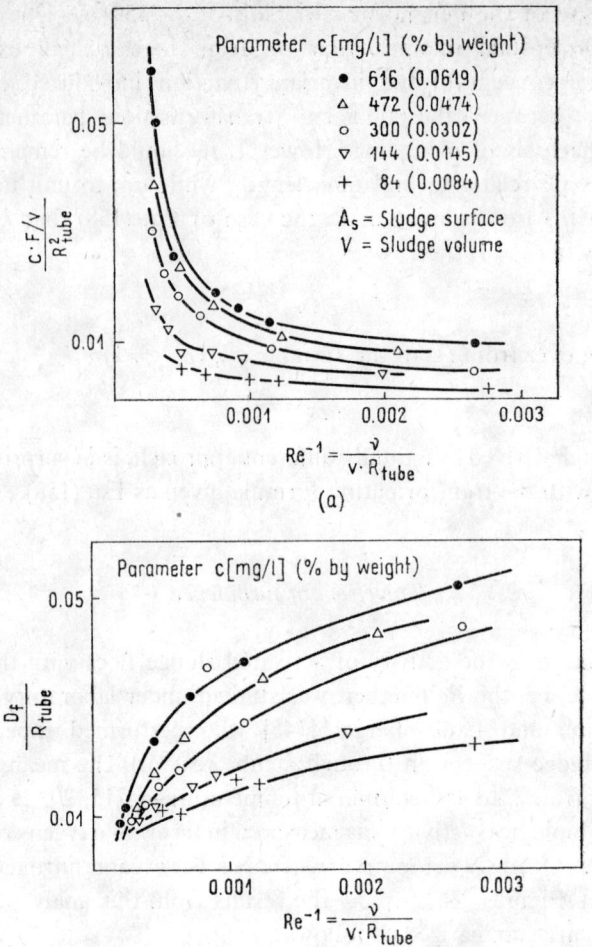

Figs 26(a) and *(b)*. Oxygen consumption of activated sludge vs the *Re* number

(d) Oxygen transport towards the cell surfaces may prove a factor limiting the rate of biological decomposition.

In one of his comprehensive studies Laubenberger [146] attempted to describe the major factors affecting the physical properties of the flocs using dimensionless relationships. By dimensional analysis he derived the following expressions

$$\frac{D_f}{R_{\text{tube}}} = f_1\left(c, \frac{vR_{\text{tube}}}{v}, \frac{vR_{\text{tube}}}{\eta_f/\varrho}, \frac{\varrho_f}{\varrho}\right) \tag{162a}$$

$$\frac{A_f}{R_{\text{tube}}} = f_2\left(c, \frac{vR_{\text{tube}}}{v}, \frac{vR_{\text{tube}}}{\eta_f/\varrho}, \frac{\varrho_f}{\varrho}\right) \tag{162b}$$

where the floc parameters are the floc density ϱ_f, the "floc viscosity" η_f, the sludge concentration c; the suspension parameters are the sludge density ϱ, the kinematic viscosity v; the flow parameter is the mean velocity v of the suspension; and the geometric dimensions are the hydraulic radius R_{tube} of the tube reactor, the mean diameter D_f of the flocs and the total surface area F_f of the flocs.

The dimensional correlation of the properties of an activated sludge sample from a treatment plant (sludge age ten days) is shown (after Laubenberger) in Fig. 27. From the above analysis it can be seen that the Re number plays a double role in forming the floc structure.

The effect of turbulence in scale-up problems was examined by Kalinske [147] in connection with activated sludge aeration basins. He emphasized the need of exercising great care in generalizing laboratory results to prototype designs if only for the effect of turbulence. It is therefore essential to define correctly the intensity and characteristic dimensions of macro- and micro-turbulence (e.g. the frequency spectrum of pulsation velocities, the average plume diameter). Kalinske demonstrated by experiment that turbulence of any given intensity is substantially simpler to realize in the model than in the prototype.

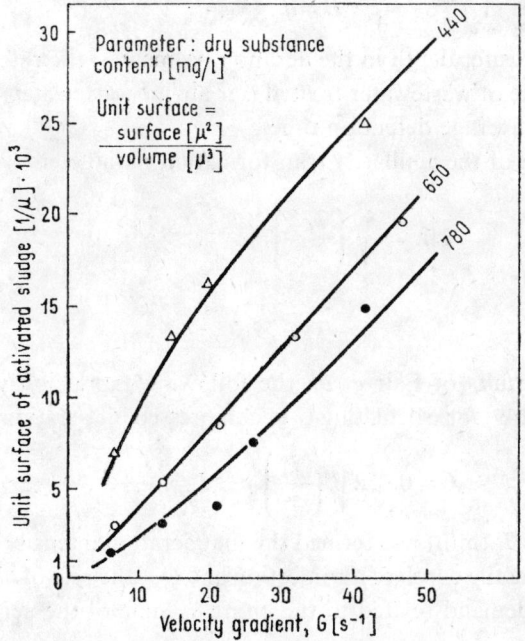

Fig. 27. Some main parameters of activated sludge systems

Kalinske also noted the importance of measurements in prototype structures to also check the applicability of the scale-up methods. From the scaling up viewpoint it is useful to have identical micro-turbulence parameters (intensity and size) in the model and prototype. To achieve this, geometrical similarity is not an essential prerequisite; under certain circumstances even conflicting requirements (e.g. the provision of baffles) may have to be satisfied.

Energy dissipation and velocity gradient

Fair et al. [148] in 1969 approached the problem of biological flocculation in aeration basins by considering energy dissipation phenomena. Using evidence obtained in plant-scale equipment they demonstrated that the efficiency of treatment can be improved by increasing the intensity of turbulence. Starting from the fundamental studies of Camp and Stein [90] they derived the following expression for calculating the energy input and average velocity gradient in compressed air aeration basins [149]:

$$G = \sqrt{\left(\frac{D}{\eta V}\right)} = \sqrt{\left(\frac{ghq_{air}}{vt}\right)} \tag{163}$$

$$D = q_{air} Q \varrho g h \tag{164}$$

where h is the water depth in the aeration basin; q_{air}, the rate of air supply per unit volume of wastewater treated (m³ air/m³ wastewater); and $t = V/Q$, the calculated average detention time.

Introduction of the similarity transformation parameters ($\lambda_v = \lambda_\varrho = \lambda_g = 1$) gives

$$\lambda_G = \sqrt{\left(\frac{\lambda \lambda_q}{\lambda_t}\right)} = \lambda^{1/2} \lambda_q^{1/2} \lambda_t^{-1/2}. \tag{165}$$

Further

$$\lambda_D = \lambda \lambda_q \lambda_Q. \tag{166}$$

Using the results of Fair et al. the following expression was suggested by Zahradka [149] for calculating G in compressed air aeration basins:

$$G = 0.052 \sqrt{\left(\frac{J_s}{v}\right)}, \quad J_s = \frac{Q_{air} h_r}{V}. \tag{167a,b}$$

The variable J_s (m/h) was termed the unit aeration intensity by Zahradka, with h_r denoting the depth of air injection and Q_{air} the rate of air supply.

The power demand related to the entire volume of the aeration basin is

$$N = \text{constant } \gamma J_s V = \text{constant } \gamma Q_{air} h_r. \tag{168}$$

Mixing

As detailed by Camp in his paper referred to previously, Christensen and Morabito as early as 1939 performed experiments at the Massachusetts Institute of Technology in checking the validity of the model law on mixing tanks. In these experiments they used geometrically similar designs differing from each other only by the scale factor $\lambda=4$ [70]. They arrived at the conclusion that the Froude Law applied with fair approximation to the flow conditions in basins both with and without mixing (stirring) devices. The peripheral speed of the mixing rotors would be reproduced successfully on the basis of the Froude Law. On the other hand, conversion on the basis of the Reynolds Law failed to produce the required similarity of the flow patterns.

Concerning the similarity and scale-up problems of mixing processes and devices there is ample information available in the professional literature, although this information is conflicting and inadequate in several respects. The research by Rushton and Oldshue *et al.* is described first [150, 151, 152]. Without embarking upon a detailed discussion of this subject, the main conclusions, principles and expressions of potential interest in scaling-up aeration facilities will only be summarized subsequently, drawing mainly on the publications of the above authors.

The conditional equations of hydrodynamic similarity of flow in mixers are governed fundamentally by the Froude, Reynolds and Weber numbers, as well as by the power number (the modified Euler number) [151]

$$Fr=\frac{n^2d}{g} \quad Re=\frac{nd^2}{v} \quad We=\frac{n^2d^3\varrho}{\sigma} \quad Eu=\frac{N}{\varrho n^3 d^5}. \qquad (169\text{a--d})$$

It should be noted that the role of the We number in scale-up problems is not yet properly understood. In geometrically similar systems Rushton and his co-workers [150] suggested the following expression for the range of fully developed turbulence:

$$Eu = KRe^aFr^b. \qquad (170)$$

In this empirical expression the constants K, a and b are experimentally derived. For their numerical value in different designs detailed guidance can be found in the literature.

In the range of high Re numbers ($Re > 10^4$), $a=0$, Eu=constant and thus

$$N = \text{constant } \varrho n^3 d^5. \qquad (171)$$

Under conditions to this equation applies the criterion of modelling is derived—as a potential alternative—from the invariance of the power number. A corresponding approach frequently adopted consists of observing the invariance of the power demand related to unit volume, i.e. N/V. In this connection it may be of interest to mention that according to Rushton [151] perhaps no theorem on mixing has led to more false conclusions than the principle of equal power per equal volume units.

It should be emphasized that even in geometrically similar systems additional special restrictions must be introduced to realize similarity of part processes under the condition $N'/V' = N''/V'' = $ constant.

For the condition of approximate similarity of the mass and heat transfer processes in mixers Rushton suggested a dimensionless power product determined empirically. In fact, he applies the familiar correlation of Gilliland–Sherwood, which in the case of mass transfer processes involves the correlation of the Sh, Re and Sc numbers (see Eq. [131]):

$$\frac{K_L d}{D} = K \left(\frac{nd^2}{v}\right)^x \left(\frac{v}{D}\right)^y. \tag{172}$$

The numerical values of K = constant, x and y can be found by plotting the $Sh Sc^{-y}$ = constant Re^x relationship. In Rushton's concept the part played by the exponent x in model–prototype conversions is an especially important one. Changing to the similarity transformation parameters (with $\lambda_D = \lambda_v = 1$) one has

$$\lambda_{K_L} = \lambda_d^{2x-1} \lambda_n^x. \tag{173}$$

Moreover, imposing the criterion $\lambda_{K_L} = 1$

$$\lambda_n = \lambda_d^{-(2x-1)/x} = \lambda_d^{-z} \qquad \frac{n'}{n''} = \left(\frac{d'}{d''}\right)^{-z} \tag{174}$$

where $z = (2x-1)/x$. In combination with Eq. (171)

$$\lambda_N = \lambda_d^{5-3z} \tag{175a}$$

and

$$\lambda_{N/V} = \lambda_d^{2-3z}. \tag{175b}$$

It can be seen that the experimental determination of x is an essential step in writing the condition equations of scale-up. The main factors influencing the magnitude of x are the geometrical design and arrangement of the mixing or aerating device. The higher the value of x, the more effective the geometric design and the more favourable the type of flow will be. Rushton distinguished three categories:

(a) It can be demonstrated that in the special case of $x = 0.75$ the invariance principle of energy per unit volume applies and thus the model can be scaled up on the basis of $N/V = $ constant.

(b) In the cases where $x > 0.75$ less energy is needed per unit volume in the larger basin (e.g. in the prototype) than in the smaller one (e.g. in the model).

(c) Finally, in the cases where $x < 0.75$ the situation is reversed with respect to (b). This implies also that at low x values scale-up may even prove uneconomical.

There is experimental evidence to show that in the case of heat and mass transfer processes model-prototype conversions based on dimensionless correlations, which take usually the form of power products, yield results that are more acceptable in practice than those obtained by analysing the invariance of individual dimensionless numbers.

Two fundamental methods of modelling mixers and mixing processes were also mentioned by Oldshue [153]: one relying on the dimensionless numbers describing dynamic similarity; the other based on the relevant correlations. In connection with the first he pointed out that in the case of mixers no adequate information is available on the possibilities and limitations of using the individual dimensionless numbers (e.g. the Fr, Re and We numbers) for realizing dynamic similarity in practical scale-up problems. The second method applied successfully, especially for modelling gas-absorption processes, involves the identification of a relevant index number which may be a dimensional quantity as well and experimentally correlating it with the basin size. The relevant index number may be the peripheral velocity, or the speed of the mixing device or aerating rotor, the rate of air flow injected per unit volume of the fluid, power uptake, etc. The observation of the ratio $N/V =$ constant is believed especially important by Oldshue for the following two reasons:

(a) The mathematical expression correlating the ratio N/V and the basin size plots in a log–log system of coordinates approximately to a straight line so that the results of experiments made at two different scales can be extended simply to the prototype (Fig. 28).

(b) Minor changes in the design and position of the mixing device or water depth hardly affect the above correlation.

In modelling fluid mixing processes, the observation of geometric similarity was considered essential by Rushton and Oldshue alike [152]. Nevertheless, the comparison of distorted systems also appears as a potential method of modelling. In one of his papers Oldshue [154] described examples of scaling up and down geometrically distorted models. By varying the ratio of the mixer diameter to the basin size and introducing distortion in this way, several relevant quantities, the power demand per unit basin volume N/V,

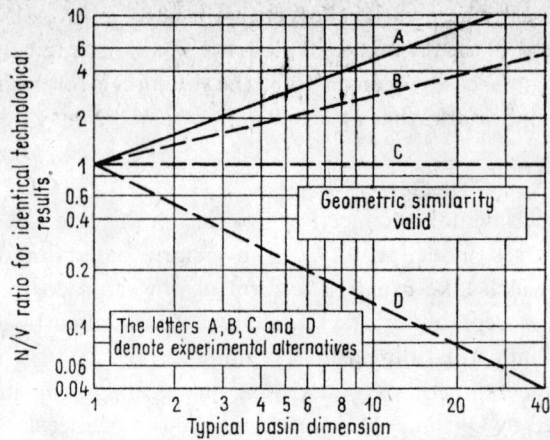

Fig. 28. The N/V ratio vs a typical basin dimension for given technological efficiency

the loading rate per unit basin volume Q/V and the peripheral velocity nd, could be maintained constant simultaneously in the two $[V''=160$ gal (606.4 l) and $V'=2500$ gal (9475 l)] systems. By violating the principle of geometric similarity he succeeded in making other quantities invariant in this way. As shown by the example quoted for scaling down, a method used in prototype scale is tested in a model for specific research purposes by operating the two systems parallel to each other (in this case $V'=2500$ and $V''=3.4$ gal).

Other relationships can also be used to advantage in designing and modelling mixers and aerators [152–155].

The power demand is the product of the discharge pumped Q and the delivery head H ($N=QH$).

The pumping capacity, or delivery rate, of a series of geometrically similar rotors is given as

$$Q \propto nd^3 \qquad \lambda_Q = \lambda_n \lambda^3 \qquad (176\text{a})$$

while the delivery head is

$$H \propto n^2 d^2 \qquad \lambda_H = \lambda_n^2 \lambda^2. \qquad (176\text{b})$$

In the case of $N=$constant power demand (see column 1 in Table 4)

$$n = d^{-5/3} \qquad d = n^{-3/5} \qquad (176\text{c})$$

so that

$$Q \propto d^{4/3} \qquad Q \propto n^{-4/5} \qquad (177\text{a,b})$$

and

$$H \propto n^{4/5} \qquad H \propto d^{-4/3}. \qquad (178\text{a,b})$$

The scale factors involved in the modelling of mixing devices have been systematically compiled in Table 4 for various invariance criteria applied

Table 4
Invariants and scale factors involved in scaling-up mixing (stirring) equipment

Representative variables				Condition equation of invariance based on dimensional number			Condition equation of invariance based on dimensional number			Condition equation of invariance based on dimensionless number		
Designation	Symbol	Dimension	Scale factor	$(n^3 \cdot d^5)' = (n^3 \cdot d^5)''$ $N' = N''$	$\left(\frac{N}{V}\right)' = \left(\frac{N}{V}\right)''$ $N_s' = N_s''$	$\left(\frac{Q}{V}\right)' = \left(\frac{Q}{V}\right)''$ $T_t' = T_t''$	$\left(\frac{Q}{F}\right)' = \left(\frac{Q}{F}\right)''$ $T_s' = T_s''$	$\left(\frac{M}{V}\right)' = \left(\frac{M}{V}\right)''$ $M_s' = M_s''$ $L_s' = L_s''$	$(n \cdot d)' = (n \cdot d)''$ $v_p' = v_p''$	$\left(\frac{n \cdot d^2}{\gamma}\right)' = \left(\frac{n \cdot d^2}{\gamma}\right)''$ $Re' = Re''$	$\left(\frac{n^2 \cdot d}{g}\right)' = \left(\frac{n^2 \cdot d}{g}\right)''$ $Fr' = Fr''$	$\left(\frac{\varrho n^2 \cdot d^3}{\sigma}\right)' = \left(\frac{\varrho n^2 \cdot d^3}{\sigma}\right)''$ $We' = We''$
				Exponents of the scale factor			Exponents of the scale factor			Exponents of the scale factor		
				1	2	3	4	5	6	7	8	9
Length	$l; d$	[L]	$\lambda; \lambda_d$	1	1	1	1	1	1	1	1	1
Surface area, A	A	[L^2]	λ_A	2	2	2	2	2	2	2	2	2
Volume	V	[L^3]	λ_V	3	3	3	3	3	3	3	3	3
Time	t	[T]	λ_t	5/3	2/3	[0]		1		2	1/2	3/2
Speed	n	[T^{-1}]	λ_n	−5/3	−2/3	[0]		−1		−2	−1/2	−3/2
Velocity	v	[LT^{-1}]	λ_v	−2/3	1/3	1		[0]		−1	1/2	−1/2
Velocity gradient	G	[T^{-1}]	λ_G	−5/3	−2/3	[0]		−1		−2	−1/2	−3/2
Acceleration	a	[LT^{-2}]	λ_a	−7/3	−1/3	1		−1		−3	[0]	−2
Discharge	Q	[L^{3}T^{-1}]	λ_Q	4/3	7/3	3		2		1	5/2	3/2
Surface loading rate	$T_s = Q/F$	[LT^{-1}]	λ_{T_s}	−2/3	1/3	1	[0]			−1	1/2	−1/2
Volumetric loading rate	$T_t = Q/V$	[T^{-1}]	λ_{T_t}	−5/3	−2/3	[0]		−1		−2	−1/2	−3/2
Force	P	[MLT^{-2}]	λ_P	2/3	8/3	4		2		[0]	3	1
Pressure	p	[ML^{-1}T^{-2}]	λ_p	−4/3	2/3	2	[0]			−2	1	−1
Torque	M	[ML^{2}T^{-2}]	λ_M	5/3	11/3	5		3		1	4	2
Unit torque	$M_s = V/M$	[ML^{-1}T^{-2}]	λ_{M_s}	−4/3	2/3	2	[0]			−2	1	−1
Work and energy	L	[ML^{2}T^{-2}]	λ_L	5/3	11/3	5		3		1	4	2
Power	N	[ML^{2}T^{-3}]	λ_N	[0]	3	5		2		−1	7/2	1/2
Unit power	$N_s = N/V$	[ML^{-1}T^{-2}]	λ_{N_s}	−3	[0]	2		−1		−4	1/2	−5/2
Power number	E	[−]	λ_E	[0]	[0]	[0]		[0]		[0]	[0]	[0]
Reynolds number	Re	[−]	λ_{Re}	1/3	4/3	2		1		[0]	3/2	1/2
Froude number	Fr	[−]	λ_{Fr}	−7/3	−1/3	1		−1		−3	[0]	−2
Weber number	We	[−]	λ_{We}	−1/3	5/3	3		1		−1	1	[0]

Notes: 1. The tabulation contains the exponents α_i of the relationship between the scale factors λ_i of the representative variables and the scale factor λ of lengths (e.g. according to column 2 $\alpha_n = -2/3$, so that $\lambda_n = \lambda^{-2/3}$, or according to column 8—the Froude criterion—$\alpha_v = 1/2$, so that $\lambda_v = \lambda^{1/2}$).
2. The validity criteria of the relationships tabulated are: (a) Observation of geometrical similarity; (b) identical physical properties ($\lambda_\varrho = \lambda_\gamma = \lambda_\sigma = 1$) of the medium flowing in the model and prototype (single and double prime) systems; (c) validity of the following relationships:
$N \propto \varrho n^3 d^5$; $N/V \propto \varrho n^3 d^2$; $P \propto \varrho v^2 l^2 \propto n^2 d^4$;
$L \propto \varrho v^2 l^3 \propto \varrho n^2 d^5 \propto M$; $M/V \propto \varrho n^2 d^2$;
$Q = Fv \propto nd^3$.
3. Relationships and tabulations for conditions other than those above can be written in a similar manner.

in practice. From this table it can be seen that under the appropriate conditions the invariance of the power number Eu plays an important role. A comparison of the data in columns 4–6 will further reveal that the specification of invariance for the surface loading rate T_s, the unit torque M (and consequently the unit energy $L_f = L/V$) and of the peripheral velocities ($\lambda_M = \lambda_{T_s} = \lambda_v = 1$) leads to identical similarity condition equations (e.g. $\lambda_n = \lambda^{-1}$). Columns 7–9 represent the familiar invariance criteria of the Fr, Re and We numbers.

In view of the above, the question may be raised whether the identity of columns 4–6 implies or does not imply the possibility of taking simultaneously several similarity criteria into consideration. No answer of general validity can be offered to this question. The invariance criteria that may simultaneously apply and improve the accuracy of the model-prototype relationship must be determined together with the special conditions to be observed for each particular case individually.

As a first theoretical example the case will be mentioned where it is wished to include the proportions given by the power number Eu ($N \propto \varrho n^3 d^5$), the unit power $N/V \propto N/d^3$ and the peripheral velocity $v \propto nd$ of the rotor.

The following line of reasoning is adopted ($\lambda_\varrho = 1$):

$$\frac{N'}{n'^3 d'^5} = \frac{N''}{n''^3 d''^5} \qquad \frac{N'}{d'^3} \frac{d'}{n'^3 d'^3} = \frac{N''}{d''^3} \frac{d''}{n''^3 d''^3}. \tag{179}$$

If $N'/d'^3 = N''/d''^3$ then

$$\frac{d'}{n'^3 d'^3} = \frac{d''}{n''^3 d''^3} \qquad \frac{d'}{d''} = \left(\frac{n'd'}{n''d''}\right)^3. \tag{180a}$$

Therefore the ratio of the peripheral velocities is

$$\frac{v'}{v''} = \frac{n'd'}{n''d''} = \left(\frac{d'}{d''}\right)^{1/3} \qquad \lambda_v = \lambda^{1/3} \tag{180b}$$

evidently in agreement with 1/3 given in column 2 of Table 4.

The second example is concerned with the mixing time, which as will be recalled is the briefest operating time needed to attain a particular mixing effect. Practical experience has shown the mixing time to be longer in large basins than in small units. Empirical formulae are available for estimating the mixing time (t_m) in mixers of different design. Of these the Holmes formula is mentioned in the first place, which applies to turbine mixers in basins with baffle plates

$$nt_m \left(\frac{d}{D}\right)^2 \approx \text{constant} \tag{181}$$

where D is the diameter of the mixing tank.

In geometrically similar systems

$$nt_m = \text{constant} \quad \text{or} \quad \lambda_t = \lambda_n^{-1}. \tag{181a}$$

Further if $N/V = \text{constant}$ $(\lambda_n = \lambda^{-2/3})$, then

$$\lambda_{t_m} = \lambda_n^{-1} = \lambda^{2/3} \tag{181b}$$

in agreement with 2/3 given in column 2 of Table 4.

This simple relationship for t_m does not apply to all mixer designs. In geometrically similar systems with turbulent flow (fermenters) Aiba *et al.* [156] adopted as the starting basis the following expression:

$$t_m n^{2/3} d^{-1/6} = \text{constant}. \tag{182}$$

The scale factor of mixing time t_m is again found by adopting the invariance criterion $N/V = \text{constant}$:

$$\lambda_{t_m} = \lambda_n^{-2/3} \lambda_d^{1/6} = \lambda_d^{11/18} \tag{182a}$$

which, however, already implies a new criterion as compared with that given in column 3 of Table 4.

From this discussion it will be concluded that the quality of mixing is scale dependent. This seems to be supported by the investigations of Biggs [157], Fox and Gex [158], and Norwood and Metzner [159]. Based on these results Hansford and Humphrey [160] provided additional experimental evidence.

The third example related to mixing will be given with reference to Hansford and Humphrey who studied the influence of the size of the experimental equipment and of mixing intensity in continuous fermentation at low dilution rates. They found the quality of mixing to depend also on the inlet design, e.g. its arrangement in the unit. They analysed the role of the mixing time for its scale-up implications, with reference to Eqs (182) and (182a) in this context. Moreover, relying on the studies of Biggs, they correlated the quality of mixing with the concentration ratio C_m/C_a determined by tracer studies. (Here C_m is the highest, while C_a the terminal tracer concentration at the end of the mixing process in some point within the mixing tank, in the case of instantaneous tracer injection.) For mixers of different design they obtained the following correlation [160]:

$$\frac{C_m}{C_a} = \text{constant} \left(\frac{D}{d}\right)^{2/3} \left(\frac{nd^2}{v}\right)^{-1/6} \tag{183}$$

where in addition to the notations introduced before, D is the diameter of the tank.

In the case where $\lambda_{C_m} = \lambda_{C_a} = 1$ and geometric similarity is observed $(\lambda_D = \lambda_d)$, the condition equation of similarity is in agreement with the

Reynolds model law ($\lambda_n = \lambda^{-2}$, see the exponent -2 in column 7 of Table 4). Studies performed here on vertical-shaft aeration tanks by series of measurements in geometrically similar or approximately similar models using clear water have resulted in values ranging from -1.8 to 2.3 for the theoretical value $\alpha = -2$.

The three examples given are believed to illustrate clearly that in modelling mixers special scale-up criteria can also be taken into consideration. A large number of reports on practical applications can be found in the subject literature. Reference is made concerning aeration units to a chapter in the Environmental Protection Agency report edited by Boyle [161], in which Salzman and Lakin examined the role of mixing in activated sludge basins, applying Eqs (171), (176a) and (177a). By theoretical considerations they arrived at the conclusion that the pumping capacity of the aerator rotor is proportionate to four-thirds the power of the diameter. It should be emphasized, however, that the pump discharge calculated in this way means only the discharge defined in the immediate vicinity of the rotor. The full discharge circulated in the basin may be several times as high. As an additional important conclusion they established that for pumping capacity rotors of larger diameter lower speed operation is preferable. Evidently, this statement does not apply to the optimization of oxygenation performance. Also the importance of observing geometric similarity was emphasized. Model studies in this domain are considered relevant and especially reliable for the analysis of hydraulic processes and for the visual examination of the flow pattern.

The mixing of media containing floating matter (suspensions) or the process of dispersion may warrant the inclusion of additional condition equations. According to the experiments of Hobler and Zablocki (see Fejes [290]), during the slurry mixing operation the mixer speed pertaining to the beginning of complete dispersion (where all solid particles become suspended) can be modelled on the basis of the Froude Law, provided that $Re > 1000$. Other authors have given $Re > 400$ as the validity limit. In the case of aeration basins this question is of importance in defining the conditions which must be satisfied to maintain the activated sludge in suspension.

Mass transfer, oxygenation

Up to the early 1960s rather little was published on the possibility of modelling mass transfer and oxygenation processes in aeration basins. In the majority of cases the conclusions concerning scale-up were only general in nature. Recently, however, a number of papers have appeared in which information based on experimental evidence was offered.

Kalinske [162] emphasized the difficulties associated with modelling oxygenation, attributing fundamental importance to geometric similarity. In view of the impossibility of keeping the *Fr*, *Re* and *We* numbers invariant simultaneously, he suggested the correctness of applying the method widely resorted to in modelling industrial mixers and based on the power demand per unit fluid volume, to the case of aeration reactors. He objected to the use of the α_{OC} coefficient (the ratio of oxygen input rates in waste- and clear water) as the basis of modelling, since the criterion $\lambda_\alpha = 1$ is not necessarily satisfied in systems of different size. The magnitude of α_{OC} being affected by an extremely large number of factors, some not necessarily having the same sign in the model and prototype, it follows that the determination of λ_α is questionable.

Kalinske and his co-workers arrived at the conclusion that the rate of oxygen input depends on the *Fr* number [136]

$$O_t = KQ(C_S - C) \qquad (184)$$

where

$$K = mFr^n \qquad (184a)$$

$$Fr = \frac{v_p}{\sqrt{(gh_r)}} \qquad (184b)$$

where O_t is the rate of oxygen input; Q, the pumping capacity of the aerator; $C_S - C = d_t$, the oxygen deficiency; K, a dimensionless factor; m, n, experimental constants characterizing the aeration equipment and the operating conditions; h_r, the depth of rotor submergence; and v_p, the peripheral velocity.

Upon substitution and rearranging Eq. (184) assumes the form

$$\frac{O_t}{Q\alpha d_t} = mFr^n. \qquad (184c)$$

If the *Fr* number is invariant, $\lambda_m = \lambda_n = 1$ with $\lambda_{d_t} = 1$

$$\lambda_Q = \lambda_{O_t} = \lambda^{5/2}. \qquad (185)$$

As will be recalled, the rate of oxygen input O_t can also be expressed in terms of the extended mass transfer coefficient $K_L a$ as $O_t = V K_L a C_S$, so that ($\lambda_{C_S} = 1$, $\lambda_V = \lambda^3$)

$$\lambda_{O_t} = \lambda_V \lambda_{K_L a} = \lambda^3 \lambda_{K_L a}. \qquad (186)$$

As will be explained in more detail subsequently, technological considerations may often justify adherence to the criterion $\lambda_{K_L a} = 1$, just as when modelling fermentation units, in which case $\lambda_{O_t} = \lambda^3$, which evidently is not necessarily identical with Eq. (185) consequent from the model law of Froude. Equations (185) and (186) become compatible, i.e. the hydraulic and mass transfer processes are approximately similar, when using the scale factor

$$\lambda_{K_L a} = \lambda^{-1/2}. \tag{187}$$

On the other hand, Eqs (185) and (186) and the criterion $\lambda_{K_L a} = 1$ can be satisfied at the scale factor $\lambda = 1$ (prototype) alone.

Several investigators have found the extended mass transfer coefficient $K_L a = (K_L F)/V$ (F is the mass transfer interface between two phases; V, the fluid volume) to depend on the intensity of turbulence as well, and to be even an indirect measure thereof in a certain sense. To illustrate this, it is sufficient to mention the equation suggested by Danckwerts [163], for example

$$K_L = \sqrt{(Dr)} \tag{188}$$

where K_L is the mass transfer coefficient; D, the diffusion constant; and r, the renewal rate of the diffusion interface, which is actually a frequency.

Evidently r is a function of turbulence. Considering that the specific surface $a = F/V$ is also greatly influenced by turbulence, it will be realized that the magnitudes of $K_L a$ and $OC = K_L a C_S$ (where OC = oxygenation capacity) are also closely related to the intensity of turbulence.

Within the validity range of Eq. (188) the scale factor of $K_L a$ is ($\lambda_D = 1$, $\lambda_r = \lambda_t^{-1}$)

$$\lambda_{K_L} = \lambda_r^{1/2} = \lambda_t^{-1/2}. \tag{189}$$

The relations between the quantities characterizing the mass transfer and hydraulic phenomena are often written in dimensionless form as theoretical, semi-empirical or fully empirical formulae, following the commonly adopted practice in chemical engineering. Examples related to the oxygen input processes in aeration reactors will be mentioned subsequently, which may serve—within certain limits—also as the basis of determining the condition equations of modelling.

Starting from the findings of several authors, Eckenfelder and O'Connor [52] described compressed-air aeration systems by correlating the Sherwood, Reynolds and Schmidt numbers, using the measurement data given in Fig. 29:

$$\frac{K_L d_B}{D} h_r^{1/3} = \left(\frac{d_B v_B}{v}\right) \left(\frac{v}{D}\right)^{1/2} \tag{190a}$$

or

$$Sh h_r^{1/3} = Re Sc^{1/2} \tag{190b}$$

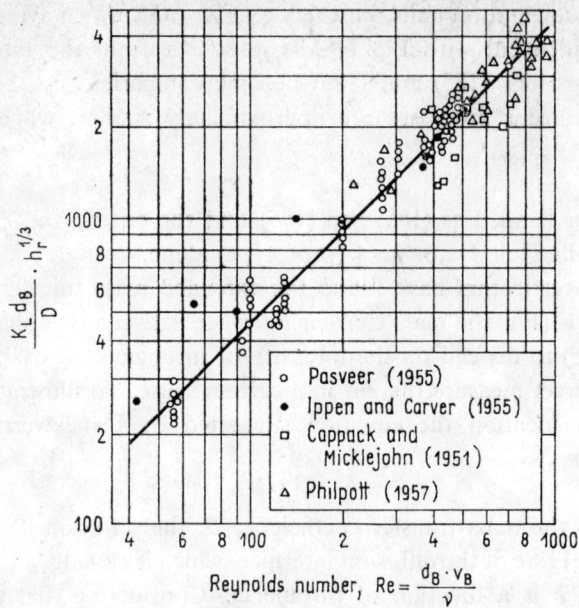

Fig. 29. Dimensionless correlation for compressed-air aeration basins with clear water

where h_r is the depth of air injection; d_B, the average bubble size; and v_B, the rising velocity of the bubbles.

The validity range of the foregoing expression is given as

$$50 < Re < 500$$
$$0.18 < h_r \text{ (m)} < 3.65.$$

The scale factor of K_L is within the validity range of Eq. (190)

$$\lambda_{K_L} = \lambda_{v_B} \lambda_{h_r}^{-1/3} \qquad (191)$$

(if $\lambda_D = \lambda_v = 1$).

Harremoës [139] adopted dimensional analysis in his study on oxygen input processes in compressed-air aeration basins, using the dimensionless numbers derived. He found the Froude number to play a dominant role, whilst the Weber number was negligible. Just as in his hydraulic studies discussed previously the characteristic velocity $(gQ_0)^{1/3}$ proved successful also in describing the oxygen transfer process. Drawing on data published by other authors, Harremoës compared the relationship in the model and prototype demonstrating the existence of an unmistakable scale effect, since the model and prototype data showed no agreement in a dimensionless plot. In this context it may be of interest to note that a different grouping of the quantities obtained by dimensional analysis, which can be accomplished in several familiar ways, would probably have resulted in relations applicable to scale-up problems. In fact, the discrepancy observed in this particular case between the model and prototype behaviour is not necessarily attributable to the scale effect alone. As an example, reference is again made to Fig. 29, where the data series plotted similarly for compressed-air aeration systems in a dimensionless system are in fairly good agreement.

Kiiskien [164] tested different air diffusion designs for his oxygen input performance by comparing model and prototype equipment using the tracer method for measuring oxygenation. The problem of scaling-up was left unmentioned.

In his paper Maise [140] suggested a dimensionless expression of the form of a power product for modelling oxygen absorption phenomena, just as for describing hydraulic conditions. He adopted (after Johnstone and Thring) the dimensionless mass transfer number

$$\pi = \frac{O_t}{DdC_s} \qquad (192)$$

as the basis of discussion. In this equation O_t is the rate of oxygen input (absorption); d, the characteristic length, the rotor diameter; and C_s, the saturation concentration of O_2.

The relation suggested for determining the condition equation of scale-up is in terms of the Re, Fr and We numbers

$$\frac{O_t}{DdC_s} = \text{constant} \left(\frac{nd^2}{v}\right)^a \left(\frac{n^2 d}{g}\right)^b \left(\frac{n^2 d^3 \varrho}{\sigma}\right)^c \qquad (193)$$

where σ is the surface tension.

As in the case of Eq. (156), the author has again omitted to publish measurement data or numerical values for the constant and the exponents a, b and c. Instead he referred to the paper of Kataoka and Miyauchi [165] in which corresponding relations are given on gas absorption by fluids in open turbulent flow. Oldshue [166] called attention to the fact that the condition equations of modelling may be affected also by the driving force of mass transfer, namely the differential concentration or pressure. It can be seen that the partial pressure of oxygen assumes different values in model and prototype aeration reactors. Consequently, the rate of oxygen input will also differ in the two. In plant-practice the rate of mass transfer is controlled by the product of $K_L a$ and the oxygen deficit $(C_S - C)$. Practical experience has shown the advisability of relating the data to the conditions $C = 0$, $\lambda_{C_s} \approx 1$, although the possibility of adopting other reference conditions must not be excluded.

Of the papers published on scale-up problems associated with aerator reactors, the report by Jackson and Collins [167] on their studies on Venturi-type aeration devices is of interest. They compared two geometrically similar ($\lambda=5$) Venturi aerators (intakes 3.8 and 3/4 in., throat diameters 1.25 and 1/4 in., respectively) for their hydraulic, oxygen input and economic parameters. Practical considerations prompted them to adopt the same flow velocities ($\lambda_v=1$) in the two devices in order to ensure adequate air entrainment in the model and prototype alike. The scale factor of discharges thus became $\lambda_Q=\lambda^2=25$. To obtain different hydraulic conditions they varied the flow passing, i.e. the velocity in the throat between wide limits. Measurements on several experimental runs resulted in the following scale factors for the principal quantities:

the rate Q_{air} or air entrainment: $\lambda_{Qair}=27$,

the differential pressure ΔP
  without air entrainment: $\lambda_{\Delta P}=0.9$,
  with air entrainment: $\lambda_{\Delta P}=1.4$,

the oxygen saturation concentration: $\lambda_{Cs}\cong 1$,

the extended mass transfer coefficient $K_L A = K_L a V$ (m³/h): $\lambda_{K_L A}=9$,

the unit oxygen input O_{tf} (kg O₂/kWh): $\lambda_{O_{tf}}=0.26$.

It can be seen that in the case of $\lambda_Q=\lambda^2$ also $\lambda_{Qair}=\lambda^2$ (specifically, $\lambda_{Qair}=\lambda^{2.05}$). Although the $K_L A$ values measured in the prototype were almost nine times as high as in the model the O_{tf} parameter (kg O₂/kWh) measuring the economics of oxygen input were superior in the model. This was interpreted by Jackson and Collins to imply the advantages of using under plant conditions several Venturi aerators of smaller size, rather than a reduced number of larger units. They also found that although lower $K_L A$ values pertained to lower throat velocities, the corresponding O_{tf} values were higher. Also, the relation between the $K_L A$ coefficient and the water discharge Q passing was found to be a linear one.

From the measurement data Jackson and Collins attempted to derive the prevailing model law by examining the invariance of the *Fr, Re, Eu, We* and *Sc* numbers. The scaling up condition equations proved impossible to write in terms of conventional dimensionless numbers. The paper thus failed substantially in deriving the condition equations of similarity.

I have used the data published by Jackson and Collins in an effort to establish the modelling relationships, adopting the concept of the invariant function as the starting point [168, 169]. The results of this analysis may be summarized as follows: the fundamental hydraulic equation of Venturi meters is familiar and the energy conditions are clearly understood. On the basis of the discharge equation the relevant dimensionless number

$$\frac{(Q/F)^2}{gh\mu^2} = \frac{v^2}{gh\mu^2} = \text{constant} \tag{194}$$

which is the *Fr* number in a form modified by the discharge coefficient μ. The condition equation of scale-up ($\lambda_g = \lambda_\gamma = 1$) is

$$\lambda_v = \lambda_\mu \lambda_h^{1/2} = \lambda_\mu \lambda_{\Delta p}^{1/2}. \tag{195}$$

Further

$$\lambda_Q = \lambda_\mu \lambda^2 \lambda_h^{1/2} = \lambda_\mu \lambda^2 \lambda_{\Delta p}^{1/2}. \tag{196}$$

On substitution of the data applying to this particular case one obtains for the devices without (1) and with (2) air entrainment ($\lambda_v = 1$; $\lambda_Q = \lambda^2$)

$$\lambda_\mu = \lambda_{\Delta p}^{-1/2} \tag{197}$$

and

$$\lambda_{\mu 1} = 1.05 \qquad \lambda_{\mu 2} = 0.845.$$

Consequently, in the case without air entrainment the scale factor of the discharge coefficient λ_μ is approximately unity. It is of interest that $\lambda \neq \lambda_h$ (and $\lambda^2 \cdot \lambda_h^{1/2} \neq \lambda^{5/2}$); geometric similarity does not therefore apply to the piezometric head; the model is distorted in this respect. Evidently the *Re* number cannot be maintained invariant ($Re' \neq Re''$), or in the case of $\lambda_v = \lambda_\gamma = 1$ one has $\lambda_{Re} = \lambda$, thus $\lambda_{Re} = 5$.

The analysis of the conversion of the variables characterizing the oxygen input leads to some highly pertinent implications. The relation given by Jackson and Collins is adopted as the starting basis:

$$O_{tf} = \text{constant} \frac{K_L A C_S}{\Delta p Q}. \tag{198}$$

Jackson and Collins failed to recognize that Eq. (198) can be used to write the relation between the scale factors. In accord with the concept of the invariant function

$$\lambda_{O_{tf}} = \lambda_{K_L A} \lambda_{C_S} \lambda_{\Delta p}^{-1} \lambda_Q^{-1}. \tag{199}$$

The validity of this expression is borne out by the measurement data published. Substituting the values $\lambda_{K_L A} = 9$; $\lambda_{C_S} \approx 1$; $\lambda_{\Delta p} = 1.4$ and $\lambda_Q = 25$, one obtains $\lambda_{O_{tf}} = 0.257$, which is in surprisingly good agreement with the value 0.26 derived from the measurement data. Consequently, this may be regarded an experimental verification of Eq. (199). Of interest is the case $\lambda_{O_{tf}} = 1$, where the quantity O_{tf} (kg O_2/kWh) is the same in the model and prototype. The condition is

$$\lambda_{K_L A} = \lambda_{\Delta p} \lambda_Q \lambda_C^{-1}. \tag{200}$$

Kalinske [162] was also of the opinion that the condition $\lambda_{O_{tf}} \approx 1$ can be satisfied in practice. In the example quoted, the equipment operated at the plant size had a capacity of 50–75 h.p., whilst that in the model was of 1 h.p. Reference is further made to Eq. (184) suggested by Kalinske and his co-workers, where $O_t = OCV = K_L a V C_S = K_L A C_S$ is directly proportional to the discharge Q, in that $K_L A C_S = KQ(C_S - C)$ and in the case of $C = O$, $K_L A = KQ$. Consequently, in the oxygen transfer process under consideration the dimensionless number $K = K_L A/Q$—the ratio of the rates defined by the oxygen transport and the fluid transport—plays a role of fundamental importance.

The relations $K_L A = KQ$ and $Q \propto nd^3$ in Eq. (176a) were adopted by Quigley and Boyle, for example, as the basis of model studies on the reoxygenation of streams by means of turbines [170]. The specific recommendations for converting the results to full-scale equipment lend special interest to their paper.

The scale factor λ_A of the phase interface A can also be estimated with a fair degree of accuracy. Jackson and Collins [167] put forward the relation $\lambda_{Q_{air}} \approx \lambda_Q = \lambda^2 = \lambda_A$ under the condition $\lambda_v = 1$ for this calculation, which also follows from the criterion of geometric similarity. This problem can also be approached from an entirely different aspect. Start from the relation of Danckwerts [163] given as Eq. (188), already used in deriving Eq. (189). In this particular case

$$\lambda_{K_L} = \lambda^{-1/2} = 0.447 \tag{201}$$

since $\lambda_v = 1$ consequent from $\lambda = \lambda_t$. Thus $(\lambda_{K_L A} = 9)$

$$\lambda_A = \frac{\lambda_{K_L A}}{\lambda_{K_L}} = 20.13 = \lambda^{1.9}. \tag{202}$$

Instead of the theoretical value 2 of the exponent, the measurement data yields 1.9, implying that the condition of geometric similarity applies with fair approximation to the phase interfaces A.

The scale factor of the coefficient $K_L a$ widely used in wastewater treatment practice is theoretically (assuming geometric similarity)

$$\lambda_{K_L a} = \lambda_{K_L} \lambda_A \lambda_V^{-1} = \lambda_{K_L} \lambda^{-1} = \lambda^{-3/2} = 0.089 \tag{203a}$$

while the measurement data give

$$\lambda_{K_L a} = \lambda_{K_L} \lambda_A \lambda^{-3} = 0.072. \tag{203b}$$

In biological treatment plants, where observation of the condition $\lambda_{K_L a} = 1$ is considered desirable, the Venturi aerator examined fails to perform satisfactorily, although the oxygen input rate is not objectionable.

It can be seen that the scale-up relations derived theoretically for Venturi-type aerators are substantiated by experimental evidence.

Recently Schmidtke [171] described remarkable research work on modelling vertical-surface aerators. Schmidtke together with myself submitted a comprehensive paper [172] to the 1976 International Conference of the IAWPR, reviewing some advances towards the modelling of vertical-shaft surface aerators on the basis of similarity theory. The joint study was based on the treatise presented at the 1966 conference of the IAWPR [173] to which more detailed reference will be made in subsequent chapters. The results to be described here are from references [171] and [172] and supplemented with a few considerations made during the preparation of this book.

Schmidtke used tap water for the aeration tests in three geometrically similar models (5.13, 19.7 and 113 l) and in a pilot unit (607 l) geometrically similar to the former. The experimental conditions were as follows: $T \approx 20\,°C$, $h_r/d_r = 0.202$, $d/h = 2.0$, $d/d_r = 4.0$ (where h_r is the rotor submergence, d_r the rotor diameter, d the width of the aeration tank, h the water depth). Schmidtke adopted the Fr number to characterize the hydraulic conditions and operated the three models M_1, M_2 and M_3 in the range

$$0.084 \leq Fr = \frac{n^2 d_r}{g} \leq 0.342 \tag{204a}$$

and the pilot unit P_1 in the range

$$0.019 \leq Fr = \frac{n^2 d_r}{g} \leq 0.132. \tag{204b}$$

The problem consisted of determining the scale-up condition equations of the oxygen input process characterized by the extended mass transfer coefficient $K_L a$. By analysing the measurement data they succeeded in writing the $K_L a/n = f(Fr)$ relation, which took in this particular case the dimensionless form

$$\frac{K_L a}{n} = mFr + b. \tag{205}$$

The constants m and b assumed different values in the four units.

In order to generalize the measurement data it was advantageous to derive a relationship equally applicable to the four different units. Under the particular experimental conditions the following empirical expression of the form of a power product was obtained:

$$K_L a = 0.98 \times 10^{-8} n^{2.41} d_r^{1.55} \tag{206}$$

if

$K_L a$ (min^{-1}); n (min^{-1}); d_r (cm).

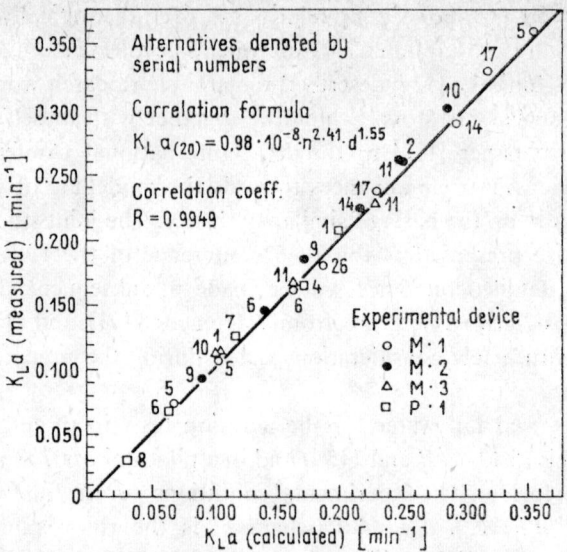

Fig. 30. Measured and calculated $K_L a$ values in geometrically similar models of different size

Schmidtke considered in more detail the case, where in analogy to the modelling of some fermentation reactors, adherence to the $\lambda_{K_L a} = 1$ criterion is advantageous. In this case the scale-up condition becomes from Eq. (206)

$$n'' = n' \left(\frac{d'_r}{d''_r} \right)^{0.65} \quad \text{and} \quad \lambda_n = \lambda_{d_r}^{-0.65} \tag{207}$$

(see Fig. 30).

Interesting conclusions may be arrived at by comparing the efforts of several authors. As the first example I will mention the studies of Robertson referred to before [132]. By assuming constant power demand per unit fluid volume, Eq. (149b) can be derived, which is in strikingly good agreement with Schmidtke's Eq. (207), in that $z = 0.65 \approx 0.67$. It should be noted that according to Connolly and Winter [155] scale-up on the basis of $N/V =$ constant may result in over-dimensioned aeration tanks. Kalinske [174], on the other hand, noted that the power demand is also significantly influenced by the geometric design of the tank and the rotor, although this effect is not experienced in geometrically similar systems.

Equation (174) quoted after Rushton implies that $z = 0.65$ and $x = 0.74$ are corresponding pairs of values. This is in agreement with $x = 0.75$ following from Rushton's theory, demonstrating again that Schmidtke's conversion method ensures simultaneously invariance of the power demand per unit volume as well. It is worth noting in this context that adherence to this criterion must not be regarded the exclusive and only applicable way of modelling. Following Rushton's approach itself it can be demonstrated that the ranges $x<0.75$ and $x>0.75$ are also well defined from the scaling-up viewpoint. Other familiar model laws can also be adopted to express the exponent z. Invariance of the Fr, Re and We numbers yields $z_{Fr}=0.5$, $z_{Re}=2.0$ and $z_{We}=1.5$. The value $z=0.65$ is remarkably close to the exponent resulting from the Froude Law. The difference in numerical value is, of course, logical, since the starting assumptions are also different.

These comparisons suggest that under certain conditions geometrically dissimilar systems can also be compared with each other and the scale-up relations can be extended to such systems as well. To illustrate this point compare the results of Schmidtke [171] with those of Cleasby and Baumann [175]. In the two series of experiments neither the aeration tanks nor the rotors were geometrically similar. The data can, nevertheless, be compared by examining only the values corresponding to the ratio $h/d=0.202$.

Examination of the data published by Cleasby and Baumann yields under the foregoing restriction the following result:

$$K_L a = 6 \times 10^{-6} v_p^{2.36} V^{-0.25} \tag{208}$$

where v_p is the peripheral velocity of the aeration rotor (cm/s); and V, the net volume of the tank (l). Since v_p constant$_1 n d_r$ and $V = $ constant$_2 d_r^3$, upon substitution one has

$$K_L a = \text{constant } n^{2.36} d_r^{1.52}. \tag{209}$$

With $\lambda_{K_L a}=1$ the scale-up relation obtained is again Eq. (207) with fair approximation ($z=1.52/2.36=0.64\approx0.65$). However, attention is called to the fact that the criterion of agreement is $h_r/d_r=0.202$ and that $\lambda_{K_L a}=1$ in the geometrically dissimilar systems. Moreover, Schmidtke claimed the rapid renewal of the gas-fluid interface, i.e. that violent turbulence also plays a role in obtaining closely identical results, regardless of geometric similarity. This fact may be interpreted as a special case of the self-modelling range. In the more general case, where the condition $\lambda_{K_L a}=1$ is not specified, scale-up relations differing to a certain extent from Eqs (206), (208) and (209) are obtained

$$\lambda_{K_L a} = \lambda_n^{2.41} \lambda_{d_r}^{1.55} \tag{210}$$

$$\lambda_{K_L a} = \lambda_{v_p}^{2.36} \lambda_V^{-0.25} \tag{211}$$

$$\lambda_{K_L a} = \lambda_n^{2.36} \lambda_{d_r}^{1.52}. \tag{212}$$

As an additional example of comparing the results found independently by different authors, the paper of von der Emde [176] is quoted. By processing the data of a number of test series he derived the empirical formula (see Fig. 31)

$$O_t = \text{constant}_1 d_r^2 v_p^3. \tag{213}$$

Since $O_t = OCV = K_L a C_S V$ and the basin volume V is proportionate to d_r^3

$$K_L a = \text{constant}_2 d_r^2 v_p^3 V^{-1} = \text{constant}_3 n^3 d_r^2 \tag{214}$$

whence

$$\lambda_{K_L a} = \lambda_n^3 \lambda_{d_r}^2 \tag{215}$$

and in the case of $\lambda_{K_L a} = 1$ the scale-up relation of Eq. (149b) is again obtained, so that $z = 2/3 = 0.67$. The ratio N/V is also invariant in this case. As will be noted, adherence to the ratio $h_r/d_r = 0.202$ mentioned repeatedly before has not been specified in this instance.

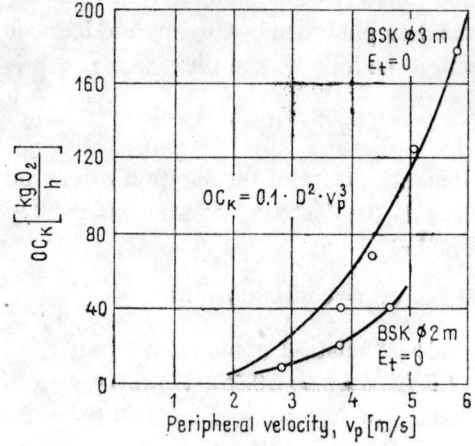

$OC_K = a(1 + b \cdot E_t) D^n \cdot v_p^m \cdot \alpha$

	D	a	b	n	m
BSK	2 m	0.1	$2.9 \cdot 10^{-2}$	2	3
BSK	3 m	0.1	$3.3 \cdot 10^{-2}$	2	3
SIMC	2.3	0.054	$3.1 \cdot 10^{-2}$	2	3
SIMC	3.6	0.054	$1.9 \cdot 10^{-2}$	2	3

BSK: $V = 1200$ m^3
SIMCAR: $V = 110 \cdot 180 \cdot 600$ m^3
E_t = Submergence
D = Rotor diameter

Fig. 31. Effect of peripheral velocity and rotor diameter on rate of oxygen transfer

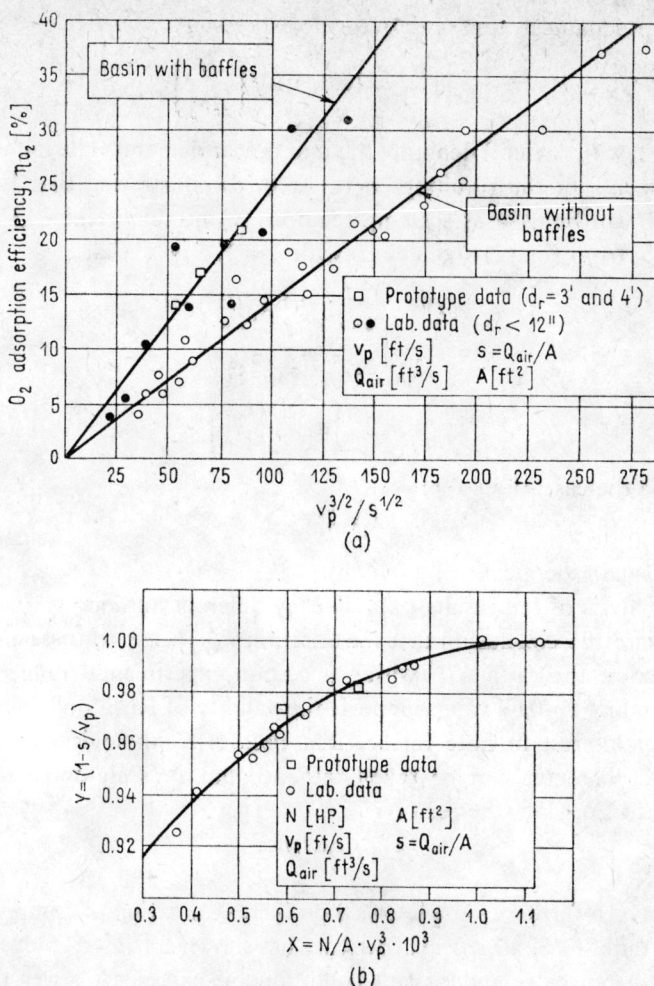

Figs 32(a) and *(b)*. Comparison of laboratory and prototype data on aeration systems

An example of modelling a turbine aeration reactor combined with air injection was published by Kalinske [177], who derived from the results of laboratory model tests an empirical formula correlating the efficiency of oxygen absorption η_{O_2} and the operating parameters of the aerator (Fig. 32/a)

$$\eta_{O_2} = \text{constant}\, \frac{v_p^{3/2}}{s^{1/2}} \tag{216}$$

where v_p is the peripheral velocity of the rotor; s, Q_{air}/A; Q_{air}, the rate of air supply; and A, the surface area of the mixing device (aerator). By determining the power demand Kalinske derived the expression

$$y = \left(1 - \frac{s}{v_p}\right) = f(x) = f\left(\frac{N}{Av_p^3}\right) \tag{217}$$

(Fig. 32b), with N (h.p.) denoting the unit power demand. The data of plant measurements fit the resulting curves well, demonstrating that Eqs (216) and (217) can be used as scale-up relations. Continuing Kalinske's line of reasoning, from Eq. (216) one finds with $\lambda_\eta = 1$, $\lambda_A = \lambda^2$ that

$$\lambda_{v_p} = \lambda_s^{1/3} = \lambda_{Q_{air}}^{1/3} \lambda_A^{-1/3} = \lambda_{Q_{air}}^{1/3} \lambda^{-2/3} \tag{218}$$

$$\lambda_{Q_{air}} = \lambda_{v_p}^3 \lambda^2 = \lambda_n^3 \lambda^5 \tag{219a}$$

$$\lambda_{q_{air}} = \lambda_n^3 \lambda^2 \quad \left(q_{air} = \frac{Q_{air}}{V}\right). \tag{219b}$$

Finally in the case of $\lambda_{q_{air}} = 1$

$$\lambda_n = \lambda^{-2/3}$$

which is again identical to Eq. (149b).

The analysis of the results published by different authors can be seen to substantiate the conclusion that the criterion $\lambda_{K_L A} = 1$ is consistent with the invariance of the ratio N/V within a certain experimental range and the corresponding $z = 0.67$ value supports the validity of Rushton's theory.

It is of interest to note further that besides technological parameters, economic character can be taken into account in scale-up analyses. In addition to Eq. (198) the relation

$$K_L a \propto Q_{air}^{0.8} N^{0.8} \tag{220}$$

is quoted as a further example based on the examination of compressed-air aeration tanks [153]. Practical experience has shown activated-sludge systems to operate normally under plant conditions at exponents lower than 0.8.

Several authors have stated that the results of oxygen transfer tests made in prototype size units are only acceptable and accurate enough to serve as bases of design criteria. Concerning the appraisal of economic parameters Fischerström [131] may be listed among these authors. In Zahradka's opinion [149] the performance of aerators is impossible to predict from the geometric characteristics of the aerator rotor and/or the tank, but should be determined from tests on prototype-size equipment. In this respect "prototype size" may be understood to include section models as well, where the cross-sectional dimension is identical with that of the prototype, but the length of the tank may be reduced to e.g., 1–2 m in the case of experimental aerators of the Inka or Kessener type. In his comments on Zahradka's paper Rincke [178] reported oxygen transfer experiments made in three geometrically similar, deep injection aerators of different size. Unfortunately, he presented no guidance of a form applicable in practice to the solution of scale-up problems.

In an attempt to expound the subject further, reference is made to the more recent work of Zlokarnik [179] who approached absorption phenomena by applying similarity theory, specifically dimension analysis, and whose results are expected to contribute in the future also to the solution of scale-up problems related to aeration tanks at sewage treatment plants.

Zlokarnik studied the rates of absorption in mixing vessels using clear water and aqueous salt solutions. By dimensional analysis he wrote the general form of the empirical relation describing the process:

$$(K_L a)^* = f_1[(P/Q)^*; \quad (q/V)^*; \quad \sigma^*; \quad Sc; \quad S_i^*] \tag{221}$$

where the individual dimensionless numbers are defined as

$$\begin{aligned}
(K_L a)^* &= K_L a (v/g^2)^{1/3} \\
(P/q)^* &= (P/q)[\varrho(vg)^{2/3}]^{-1} \\
(q/V)^* &= (q/V)(v/g^2)^{1/3} \\
\sigma^* &= \sigma[\varrho(v^4 g)^{1/3}]^{-1} \\
Sc &= v/D \\
S_i^* &= \text{the coalescence index} \\
&\quad \text{of the solutions}
\end{aligned} \tag{222a–f}$$

(which depends among others on the electric charge of the ions).

The notations used in the foregoing expressions are: ϱ, v and σ, the physical properties of the medium, namely density, kinematic viscosity and surface tension, respectively; g, gravitational acceleration; D, diffusion coefficient (gas-fluid); P/q, the power demand of dispersion related to unit gas supply (air discharge introduced); q/V, the gas supply related to unit volume of mixed fluid; and Sc, the Schmidt number.

By processing the measurement data graphically, Zlokarnik determined the dimensionless relations for the two main alternatives, namely clear water as mentioned before (coalescent conditions) and aqueous salt solution (non-coalescent conditions). The data and analysis on these two alternatives are shown in Figs 33a and b, indicating the empirical relations obtained and the ranges of validity. [The additional notations in Figs 33a and b are: G, the gas throughput through the interface (kg/s); q, the gas throughput (m³/s); Δc_m, the log mean concentration difference (ppm or kg/m³); H, liquid height above the stirrer (mm or m); D, the vessel diameter (mm or m).]

Zlokarnik, from his series of experiments, drew several conclusions, one

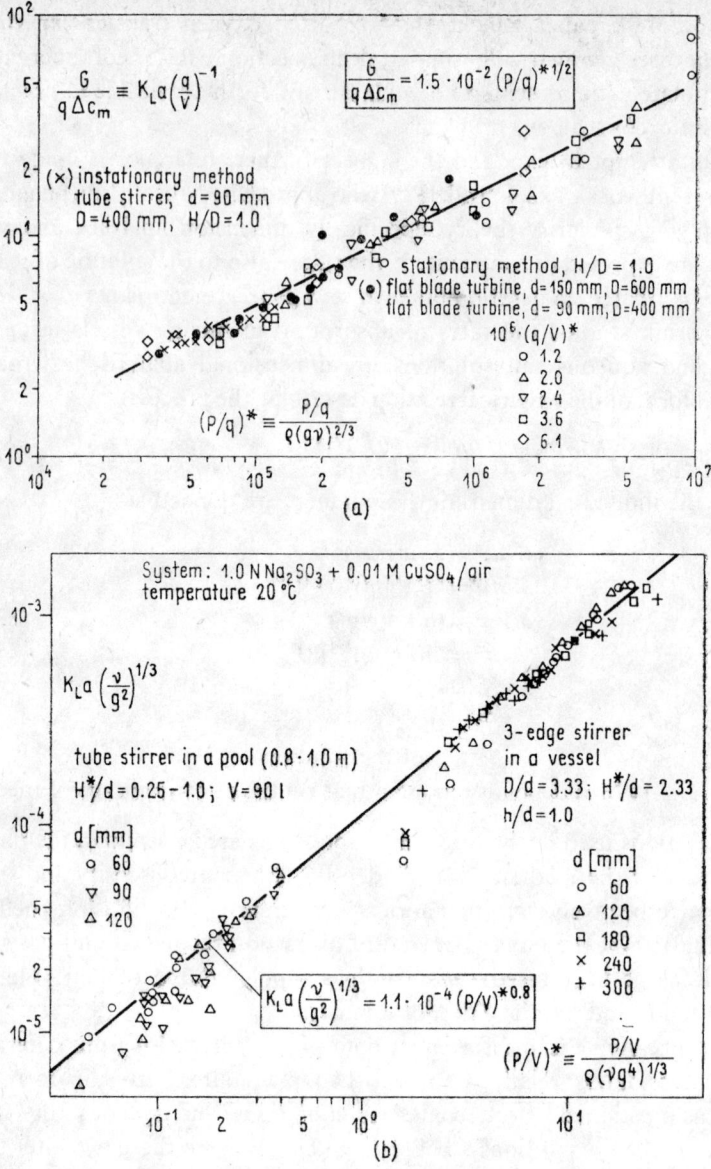

Figs 33(a) and (b). Sorption characteristic for hollow stirrers and flat blade turbines in a coalescent system. Sorption characteristic for hollow stirrers in a noncoalescent system

being that for scaling up this particular kind of absorption (non-coalescent) system the smallest tank diameter is $D=500$ mm. Once the stirrer diameter attains the value $d=180$ mm ($D=600$ mm, $V=170$ l), the dimensionless relation becomes scale independent. The reason for this is that in larger units the mass transfer across the tank water (open) surface becomes insignificant. This phenomenon may also be regarded as a typical case of the scale effect, the neglect of which is believed to explain the inconsistencies in several research results.

In another paper Zlokarnik [180] investigated the possibilities of specifically modelling the aeration systems in wastewater treatment technology, confining his attention to vertical-shaft rotors. Some of the more interesting results on modelling oxygen input (mass transfer) are reproduced below:

(a) The dimensionless relation obtained by dimensional analysis for vertical-shaft rotors is

$$\frac{G}{\Delta c d^3}\left(\frac{v}{g^2}\right)^{1/3}=f_2\left(\frac{n^2 d}{g};\ \frac{nd^2}{v};\ \frac{v}{D};\ \frac{\sigma}{\varrho(v^4 g)^{1/3}}\right) \quad (223)$$

where the individual dimensionless numbers are defined as follows:

Sorption number $\quad Y=\dfrac{G}{\Delta c d^3}\left(\dfrac{v}{g^2}\right)^{1/3}$

Froude number $\quad Fr=\dfrac{n^2 d}{g}$

Reynolds number $\quad Re=\dfrac{nd^2}{v}$ $\hfill (224\text{a--e})$

Schmidt number $\quad Sc=\dfrac{v}{D}$

Material number $\quad \sigma^*=\dfrac{\sigma}{\varrho(v^4 g)^{1/3}}=\dfrac{(Re^4 Fr)^{1/3}}{We}$.

(b) In analogy to the experimental results on stirrers—in the range of turbulent flow ($Re > 10^4$)—the author assumed the effect of the Reynolds number to be negligibly small

$$Y=f_3(Fr) \quad Sc=\text{idem} \quad \sigma^*=\text{idem}. \quad (225\text{a--c})$$

This expression may be referred to as the sorption correlation.

(c) The data processed have been plotted in Figs 34a and b, also indicating schematically the aerator system investigated. (It should be noted that I studied several rotor designs, with diameters ranging from 200 to 400 mm, reproducing the corresponding prototype at scale factors $\lambda=10$ and $\lambda=20$.)

(d) Comparing the results published by Schmidtke and Horváth [172] and the data of Zlokarnik [180] with reference to Fig. 35 it can be seen that the effect of the Reynolds number is important enough to be taken into considera-

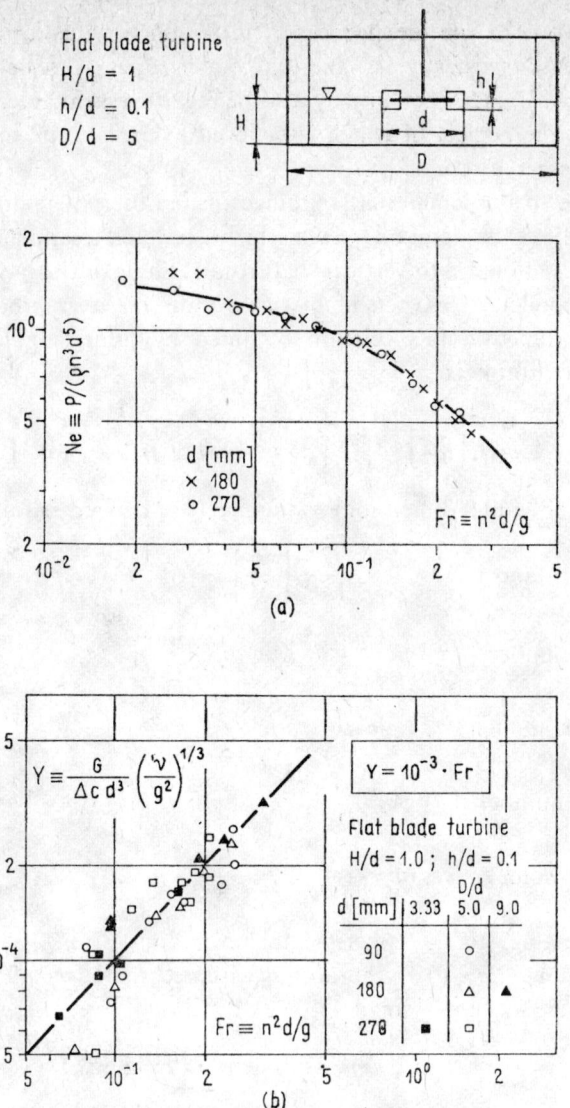

Figs 34(a) and *(b)*. Power correlation *Ne(Fr)* for a flat-blade turbine positioned at the liquid surface. Sorption correlation *Y(Fr)* for a flat-blade turbine positioned at the liquid surface

tion (see also the subsequent part on power demand and energy consumption).

(e) After corresponding reappraisal, Zlokarnik introduced the Reynolds number with the help of the Galilei number $[Ga = Re^2/Fr = (d^3g)/v^2]$, referring

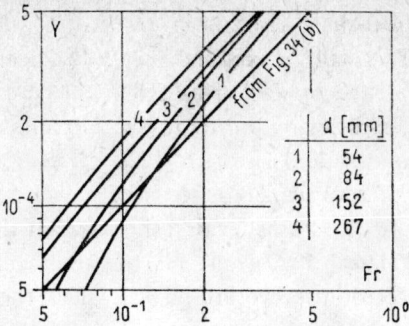

Fig. 35. Results of Schmidtke and Horváth on flat-blade turbines represented in the form $Y(Fr)$ and compared with Zlokarnik's measurements

to the investigations of Schmidtke and Horváth. The relation given by Eq. (206) can be rewritten accordingly as

$$Y = 1.41 \times 10^{-4} Fr^{1.205} Ga^{0.115} \qquad (226)$$
$$0.02 < Fr < 0.34$$
$$1.5 \times 10^9 < Ga < 2 \times 10^{11}.$$

(f) By processing the measurement data of Groot Wassink *et al.* [181] Zlokarnik arrived at the same result (see the next subsection):

$$Y = \text{constant } Fr^{0.95} Ga^{0.033}. \qquad (227)$$

The differences between Eqs (226) and (227) are believed to be justified by the differences in the experimental conditions.

For a better understanding of this subject it is necessary to consider the problem of power and energy input in greater detail. In this context reference is made to my results which will be reviewed comprehensively in Sections 3.3, 3.4 and 3.5.

Power and energy input

Conversion from one size to another of the data indicating the economic performance of aeration systems is realized to present one of the most intricate problems in the domain considered. This is primarily due to the complex, intricate nature of the phenomena involved, but the difficulties of measurement, data processing and other problems also play important roles.

In principle there are different approaches possible to this question. The method adopted by Rushton *et al.* [150] for studying mixers, and already referred to in deriving the dimensionless relation of Eq. (170), will be mentioned first. I followed the same path when studying the power and energy input of vertical-shaft aerators in terms of the *Fr* and *Re* numbers in the early 1960s. This will be expounded more in detail in Chapter 3.

The approach suggested by Zlokarnik [180] will be considered here. He introduced his study by drawing on the research results of several workers, mainly those of Schmidtke and Horváth [172].

The original idea of Zlokarnik was that the power correlation could be written in the form $Ne=f(Fr)$, neglecting the effect of the *Re* number [just as in the case of the relation $Y=f_3(Fr)$]. The measurement data on flat-blade turbines yielded the relation shown in Fig. 34a, where the parameter *d* is the rotor diameter.

The efficiency of oxygenation is often described by the total unit oxygen intake rate expressed in the form $E=G/P$ (kg O_2/kWh), where *G* is the total unit oxygen intake rate (kg O_2/h or kg O_2/s); and *P* is the power input (kW). Zlokarnik, using the economic index *E*, introduced a dimensionless number denoted by E^* taking into account the relation of the quantity considered earlier and the *Ne* and *Fr* numbers

$$E^* = \frac{Y}{NeFr^{3/2}} = \frac{Gd^{1/2}}{P\Delta c} \varrho(v^2 g^5)^{1/6}. \tag{228}$$

He found the E^* number to be a function of the *Fr* number, since his investigations have shown the *Y* and *Ne* numbers to depend also on the *Fr* number. From Eq. (228), with the condition $E^*=$ idem, similarity considerations lead to the conclusion that

$$E = \frac{G}{P} = \frac{\text{constant}}{d^{1/2}}. \tag{229}$$

Upon substitution of actual data Zlokarnik realized that the application of aeration rotors of relatively small diameter ($d \approx 1.0$ m) can only be suggested. He referred to similarity theory to support this conclusion.

In view of the fact repeatedly observed in practice that vertical-shaft aerators with rotors larger than 1.0 m can also be operated economically, at high efficiency, the above conclusions call for some critical remarks. It is suggested that the **results obtained by using the exponent 1/2 of the rotor diameter in Eq. (229) are not accurate enough and are in general inconsistent** with the measurement data of various authors. (As will be seen subsequently, the exponent in question, derived from various measurement data, is in the majority of cases lower than 0.5.) Zlokarnik himself had later recognized this fact. Equation (229) alone or in combination with considerations of similarity theory is insufficient to decide on this problem, since, for example, the exponent of *d* is obtained exclusively from the criterion of dimensional homogeneity. This, however, will be recalled to yield acceptable results only under the restriction that besides the dimensionless quantities involved in Eq. (229), there are no additional factors of major influence.

One potential way of refinement seems to involve the inclusion of the Re number (as I had indicated in a paper published in 1966 [182]). Taking the results of research by Schmidtke and Horváth into consideration, Zlokarnik revised his conclusions and made allowance for the effect of the Re number by introducing the Ga number ($Ga = Re^2/Fr$). On the basis of similar theoretical considerations he rewrote Eq. (206) into the form of Eq. (226), whence he obtained for the power intake the following dimensionless relation

$$E^* = YGa^{-0.115}Ne^{-1}Fr^{-1.5} = \frac{Gd^{0.155}}{P\Delta c} f(v, g, \sigma). \tag{230}$$

The resulting exponent of 0.155 appears to be more realistic, indicating that the economics of oxygenation do not deteriorate as rapidly as the rotor diameter is increased.

With respect to the analysis of power and energy input some recent additional advances are considered important enough to be reviewed briefly.

The subject has been dealt with in several papers by Groot Wassink et al. [181]. Rácz in his studies used, among others, my paper published in 1970 [183]. In 1978, Groot Wassink et al. examined a vertical-shaft aerator of the "Ramix" type under laboratory conditions with rotor diameters of 0.185, 0.3 and 0.4 m using a square tank. From the measurement data they derived the expression

$$Ne = \text{constant}_1 Fr^{-0.25} Re^{-0.1} \tag{231}$$

indicating that the Re number also plays a role along with the Fr number. Rewritten in the manner of Zlokarnik this becomes

$$Ne = \text{constant}_2 Fr^{-0.3} Ga^{-0.05}. \tag{232}$$

Further

$$E^* = YGa^{-0.033}Ne^{-1}Fr^{-1.5} = \frac{Gd^{0.4}}{P\Delta c} f(v, g, \sigma) \tag{233}$$

where

$$Y = \text{constant}_3 Fr^{0.95} Ga^{0.033}. \tag{234}$$

The experiments of Groot Wassink et al. are especially remarkable, because of their variation of the kinematic viscosity of the fluid (between the limits $v = 1$ to 18×10^{-6} m²/s by adding polyvinyl-pyrrolidinone). By controlling the temperature the influence of diffusivity was appraised.

By reformulating the results of Roustan [184], Zlokarnik derived for (a) flat-blade turbines and (b) pitched-blade turbines the expressions

(a) $\quad\quad\quad E^* = 2.2 \times 10^{-3} Fr^{-0.37} \quad \text{if } 0.2 < Fr < 2.0$ \hfill (235)

(b) $\quad E^* = 2.2 \times 10^{-3} Fr^{-0.43} \quad \text{if } 0.2 < Fr < 3.0.$ \hfill (236)

Finally mention is made of Bruxelmane's studies [185] on flat-blade turbine aerators with a circular tank of 1.15 m diameter in combination with rotors of 0.2, 0.245 and 0.348 m diameters. The resulting expressions are in the form adopted by Zlokarnik:

$$Ne \propto Fr^{-0.37} \quad \text{if } Fr < 0.13 \tag{237}$$

$$Ne = 0.35 Fr^{-0.5} \quad \text{if } Fr > 0.35. \tag{238}$$

In the intermediate, transition range ($0.13 < Fr < 0.35$) the Newton number was found to drop steeply from $Ne = 1.0$ to 0.57. Bruxelmane's results do not reflect the influence of the Re number.

From the considerations outlined this far the following conclusions may be arrived at:

(a) Initially it can be realized that as regards power and energy intake the results of many authors are only consistent in certain respects; differences of both theoretical and practical significance are observable.

(b) These differences may be attributed to several factors. Among these the inadequacies of measurement technique (model–prototype relations determined and checked by only one or two rather than a series of measurements), the differences in experimental conditions or even subjective factors may be mentioned. The most important factor is, however, believed to be the complexity of the phenomenon considered, preventing description by a single scale-up relation of the wide variety of aerator designs over their full range of operating conditions (just as established in connection with modelling other treatment facilities).

(c) Consequently, the validity ranges of the relations obtained should be stated positively, otherwise no comparison can be expected to yield sound results. Evidently the scale-up relations derived must be used only in the ranges that are consistent with the experimental conditions. This implies at the same time that the statement excluding in principle the possibility of modelling and scaling up in this field is not considered acceptable by myself. On the contrary, under theoretically and experimentally well-founded conditions and within reasonable limits, scale-up can be accomplished with an accuracy acceptable for practical purposes. Even if a certain percentage error is involved in these calculations—consequent also from the fact of approximate similarity—the approach is still believed to be sounder than that adopting the data determined in different-scaled systems indiscriminately, without modelling considerations, as design criteria.

Similarity of bubble movement

The hydraulic and oxygen transfer phenomena in aeration basins are greatly influenced by the manner in which the bubbles move. Consequently, it would be mistaken to neglect this effect in scale-up considerations.

In a stationary medium the rising velocity w of the bubbles can be found from familiar expressions. Considering a stationary medium, one can write the following relationships [186] for typical ranges of the Re number and the bubble diameter d_B (cm)

(a) $Re<2$ $d_B<0.015$ $Re/Fr=18$ (239)

(b) $2<Re<744$ $0.015<d_B<0.21$ $Re/Fr^{3/2}=47$ (240)

(c) $744<Re<1380$ $0.21<d_B<0.72$ $We=3.64$ (241)

(d) $Re>1380$ $d_B>0.72$ $Fr=0.51$.   (242)

Expressed in terms of the quantities describing bubble movement, the dimensionless numbers are

$$Fr=\frac{w^2}{gd_B} \qquad Re=\frac{wd_B}{v} \qquad We=\frac{w^2 d_B \varrho}{\sigma}. \qquad (243\text{a-c})$$

The dimensionless relationship describing the rising velocity w is a function of the bubble diameter d_B. Considering this relationship as an invariant function, the scale-up criteria applying to the various ranges can be written on the basis of Eqs (239)–(242).

In wastewater treatment practice the greater part of the air volume (injected or entrained) present in the aeration basin takes the form of bubbles having radii larger than $r_B=2.5$ mm, especially in the so-called medium and coarse bubble systems. Since in this range the curve $w=f(d_B)$ approaches a horizontal line, the simplification $w=f(d_B)\approx$ constant is acceptable. This is also supported by the formula suggested by Levich [187], who stated that the rising velocity is virtually unaffected by the bubble size provided that $r_B>0.1$ cm. Evidently in theoretical analyses this simplification is not necessarily allowable—see Eqs (239)–(243).

From the above considerations as well as from the results of measurements related to experiments in model and prototype scales alike one may arrive at the substantial conclusion that for analysing hydraulic conditions the rising velocity of bubbles in the model and prototype may be assumed to be approximately equal, regardless of the fact that the bubbles in the model are smaller than in the flow field of the prototype. Thus $\lambda_w \approx 1$. As will be seen in Chapter 3, this conclusion can be used in writing the scale-up criteria on hydraulic conditions. Obviously this simplification must not be understood to imply that, for example, a fine-bubble aeration system could be reproduced by a medium or coarse bubble system.

A theoretical example is finally given for the determination of the mean bubble size from considerations of similarity theory. According to the investigations by Eckenfelder and O'Connor [52] the mean bubble size is a function of the rate of air supply Q_{air} injected: $d_B \propto Q_{air}^a$. From the measurement data of Pasveer (see Horváth [291, 203a]) one finds $a = 0.10$–0.44. The approximate value of a can also be derived from theoretical considerations.

The primary characteristics of bubbles rising in a flowing medium are the diameter d_B and the rising velocity v_B. Thus in combination with the variable Q_{air} the following dimensionless number is formed [188]

$$K_5 = \frac{Q_{air}}{d_B^2 v_B}. \tag{244}$$

Assuming similar processes one obtains under the conditions K_5 and $v_B = a$ constant the relation

$$d_B = \text{constant } Q_{air}^{0.5}. \tag{244a}$$

The activated-sludge aeration basin as a fermenter

As will be recalled the activated-sludge method of wastewater treatment can be dealt with in a number of ways by analogy to fermentation processes with the aeration basin regarded as a fermenter. Accordingly, reference has repeatedly been made in the literature to the fact that the principles and methods developed for scaling up fermentation reactors can also be adapted to the case of activated-sludge aeration basins. The publications of interest related to this approach will be reviewed subsequently in a comprehensive manner.

The fundamental aspects of modelling fermentation reactors are presented by Cooper et al. [189]. These authors have underlined the principles of geometric similarity and energy demand related to unit volume. From measurement data they succeeded in demonstrating that

$$K_L a \propto \left(\frac{N}{V}\right)^{0.95}. \tag{245}$$

In the range $Re > 10^4$ the ratio N/V can be entered on the basis of Eq. (171).

This proportionality was found to be valid in the model and prototype alike. The results of Karow et al. [190] substantiated these conclusions.

The reports on efforts at modelling fermentation systems were reviewed by Bartholomew [191] who demonstrated by experiment that the exponent 0.95 in Eq. (245) actually depends on the size and design of the fermentation reactors. Thus for a small (2 gal/7.58 l) reactor he found the value 0.95, while for a larger unit (110 gal) the exponent was 0.67. In the prototype fermenters (6000–12,000 gal) the values ranged from 0.50 to 0.33, depending on whether the energy introduced was controlled by the speed of the stirrer or by changing at $n=$ constant the number of the blades. From this Bartholomew drew the conclusion that the energy related to unit volume does not in general provide a satisfactory method for scaling-up fermentation reactors. This statement is in agreement with the concept of Rushton mentioned before [151], which Bartholomew himself acknowledged as one of the potential scale-up methods adapted to fermenters.

From his original investigations Bartholomew further demonstrated that the scale-up method should preferably be based instead on the mass transfer characteristics. In this respect he emphasized the decisive role of the extended mass transfer coefficient $K_L a$, for example. As a practical approach $\lambda_{K_L a} = 1$ may be specified as the scale-up criterion. To support this statement he derived the relation between fermenters of 2 and 110, and of 6000 and 12,000 gal. volume, showing the result of scaling-up to be perfect. Special importance is attributed to this method by the fact that scale-up acceptable for all practical purposes can be achieved without the need of analysing in detail the complicated processes of fermentation. Of the mass transfer processes Bartholomew suggested adopting the rate of oxygenation as the starting basis. At the same time it is considered important to ensure identical initial and boundary conditions in modelling systems of different size.

In another comprehensive paper Finn [192] considered the physical and physico-chemical process of stirring and aeration with special regard to the possibilities of modelling. He pointed out that if air is injected into the basin during the process of stirring, the power demand must also be taken into account when writing empirical relations of the $K_L a \propto (N/V)^a$ type. As an additional relevant expression the relation $K_L a \propto Q_{air}^b$ should also be determined simultaneously. The measurements of Cooper et al. [189] have shown $b = 0.67$, while for laboratory fermenters values ranging from 0.4 to 0.5 Hixon and Gaden [193] reported for b values between 0.33 and 0.82 under particular experimental conditions. This again implies that the magnitude of b—just as that of a—is influenced by the design and size of the experimental equipment, in general by the initial and boundary conditions, which it would be a mistake to neglect in model to prototype conversions.

For higher efficiency of oxygen transfer Finn preferred fermenters extended in the vertical direction, even for experimental purposes. As a demonstrative example he compared two designs, assuming that the task was to double the volume of a particular experimental fermenter ($\lambda_V=2$):

(a) One scaled up by observing geometric similarity, with dimensions enlarged according to the scale factor

$$\lambda=\sqrt[3]{\lambda_V}=\sqrt[3]{2}.$$

(b) The other increased by violating the principle of geometric similarity, with the vertical dimension of the fermenter doubled, while the other dimensions remain unchanged, for example.

The measurements of Cooper et al. [189] have shown for this latter solution of modelling a scale factor $\lambda_{K_L a}=1.4$. The scale factors of the other relevant variables are given in Table 5 for the alternatives (a) and (b) [192].

Table 5

Comparison of fermenter scale-up alternatives

Variable	Alternative	
	(a)	(b)
$K_L a$	$\lambda_{K_L a}=1.0$	$\lambda_{K_L a}=1.4$
Total energy input	$\lambda_p=2.0$	$\lambda_p=2.0$
Q_{air}	$\lambda_{Q_{air}}=1.6$	$\lambda_{Q_{air}}=1.0$
Efficiency of oxygen transfer	$\lambda_{\eta_o}=1.26$	$\lambda_{\eta_o}=2.8$

The tall fermenter according to alternative (b) will be seen to require a lower air supply than the one scaled up according to (a). This is accompanied by an improved efficiency of oxygen transfer.

Oldshue [166, 194] reported in a number of papers on experience gained with scaling-up fermenters. Of these the paper published in 1966 is of special interest in which attention is focused on the importance of the "partial modelling" method. In fact, owing to the complex nature of fermentation phenomena, scale-up is generally confined to part processes only—within a particular unit at a given instant. In this way the scale-up criteria predomi-

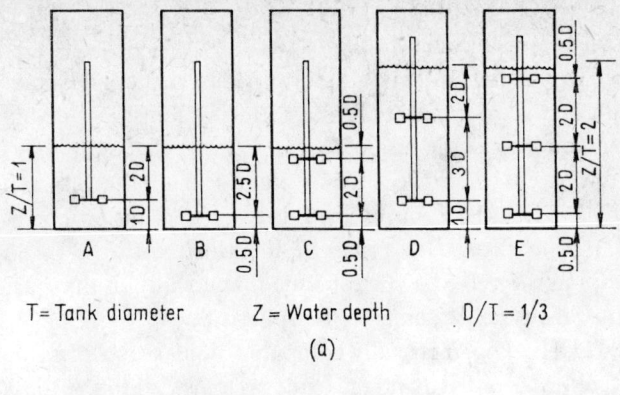

T = Tank diameter   Z = Water depth   D/T = 1/3

(a)

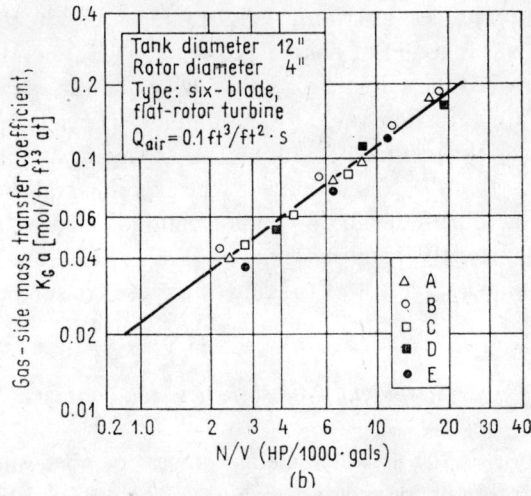

(b)

Figs 36(a) and *(b)*. Unit power consumption and the mass transfer coefficient related for fermenters of different design

nantly characterizing a particular part-process can more readily be taken into account. The invariance conditions applying to the various part-processes may accordingly differ from each other partly or entirely. (Reference is made here to the relations compiled in Table 4.)

Another inference which is also of fundamental importance in the case of activated-sludge aeration basins drawn by Oldshue concerns the relation between the mass transfer coefficients and the power demand related to unit volume. Oldshue succeeded in demonstrating (see Figs 36a and b) that in a broad range of fermenter dimensions, identical mass transfer coefficients are obtained while the ratio N/V remains invariant. (Refer to

Rushton's $x=0.75$ condition as well as to the works of Schmidtke and others previously mentioned.)

As noted by Sherwood and Pigford [195] the actual value of the Sh number can be fixed for cells in fermentation systems. For yeast production under given conditions the value $K_L d_c/D = 2$ was derived both theoretically and experimentally (where the cell diameter $d_c = 5$ μm the diffusion coefficient of oxygen $D = 1.8 \times 10^{-5}$ cm²/s and the mass transfer coefficient $K_L = 0.072$ cm/s). The ratio D/K_L may be identified with the thickness of the film enveloping the cell and corresponds numerically in this particular case to the radius of the cell assumed to be spherical.

Strom et al. [196] geometrically compared similar and dissimilar fermenters of different size (stirred and unstirred systems), adopting the $K_L a$ and the sulphite oxidation process as criteria. It is of interest to note that—although the $K_L a$ value is a function of reactor geometry—a close correlation could be observed between the sulphite oxidation value and the efficiency. Beyond a certain sulphite oxidation value (>150 mmol O_2/l per hour) the type of the fermenter and the method of aeration had no significant influence on the capacity of the system. In connection with the experimental checks on scale-up methods of aerobic fermentation units mention is made of the papers by Wegrich and Shurter [197] and Jensen et al. [198]. The latter analysed scale-up based on the measurement of the rate of oxygen consumption.

Description by dimensionless relations

The description of the activated sludge process of wastewater treatment by dimensionless numbers and relations was analysed in detail—as in the case of trickling filter systems—by Tuček et al. [126, 199]. By dimensional analysis they derived three relevant dimensionless numbers:

$$\pi_1 = \frac{C_0 - C_2}{C_0} = \frac{\Delta C}{C_0} \qquad \pi_2 = k_1 t \qquad \pi_3 = \frac{C_2}{X_1} \qquad (246\text{ac})$$

and

$$\pi_{1R} = \frac{C_{0R} - C_2}{C_{0R}} = \frac{\Delta C}{C_{0R}} \qquad \pi_{2R} = k_1 t_R \qquad \pi_{3R} = \frac{C_R}{X_R} \qquad (247\text{a-c})$$

where C_0, C_{0R}, C_2 are the substrate concentrations in terms of the BOD_5 in the inflow to the aeration basin and in the effluent from the secondary settling tank, respectively; k_1, the first-order reaction rate constant ($k_{T°C} = 0.2 \times 1.047_T^{-20}$); t, t_R, the calculated mean retention time; and X_1, X_{1R}, sludge concentration at the inlet to the aeration basin. The subscript R refers to characteristics altered by recirculation. The quantities π_1 and π_2 correspond substantially to the dimensionless numbers (π_1 and π_2) derived from a similar analysis of trickling filters.

Tuček *et al.*, having processed a number of measurement data, derived the following dimensionless relations as basic scale-up equations:

for aeration basins with plug flow

$$\pi_{1R}=\frac{\pi_{2R}}{a+1.03\pi_{2R}} \qquad a=1.9\times10^{-2}\pi_{3R}+3\times10^{-3} \qquad (248\text{a,b})$$

for aeration basins with complete mixing

$$\pi_1=1-\frac{a}{2\pi_2}\left[\sqrt{\left(1+\frac{4\pi_2}{a}\right)}-1\right]$$

$$a=2.1\times10^{-2}\,\pi_3+5.6\times10^{-5}. \qquad (249\text{a,b})$$

The only critical remark to be attached to this highly interesting paper is that the authors have also omitted to specify accurately the validity ranges of the relations suggested.

Sundstrom *et al.* [200] adopted the differential equations of aeration reactors with complete mixing as the starting basis for writing the relevant dimensionless numbers. The approach is therefore one by equation analysis (although the authors mentioned dimension analysis). Applying the method described by Hellums and Churchill (see Sundstrom [200]) and rewriting the balance equations characterizing the processes in dimensionless form, they suggested the following five dimensionless quantities as the principal variables:

the dimensionless retention time: t/t_c

the dimensionless substrate concentration: C/K_m

the dimensionless sludge concentration: $X/y \cdot K_m$

the dimensionless rate of sludge growth: $r_{max} \cdot t_c$

the dimensionless rate of endogenous respiration: k_e/r_{max}

where C is the substrate concentration; K_m, the Michaelis-Menten constant; y, the yield constant; r_{max}, the maximum value of growth rate; X, sludge concentration; and k_e, the rate constant of endogenous respiration.

From a detailed study of systems with and without recirculation and of the dimensionless relations describing them it was emphasized, as an essential result, that the balance equations and formulae written in dimensionless form provide a more general description of the technological processes in question. Reference is made here to my earlier investigations which produced similar results in several respects [201, 202].

Scaling-up activated sludge systems

Several papers have been published in the professional literature which aimed at the solution of scaling problems associated with the part-processes taking place in aeration reactors. Unfortunately, the number of scaling methods which encompass the entire technological process are founded on similarity theory and experimental verification is rare. The subsequent investigations provide some guidance to technological scaling, though not founded on considerations of similitude.

In their paper Eckenfelder and Cardenas [129] dealt with the methods of designing prototype units using the data of laboratory scale activated sludge experiments. Without offering any justification they declared laboratory treatment equipment to be suited to developing prototype design criteria. They further stated that for scaling-up laboratory units to prototype dimensions there must be identical hydraulic and biological conditions in both systems. The possibilities and methods, which would already require similarity considerations, are left unmentioned. The kinetics of BOD removal, the relations of oxygen demand and sludge growth are assumed to be identical in the model and prototype, without specifying the relevant criteria. (It should be noted that this philosophy is rather common in the wastewater treatment profession.) They further emphasized that the activated sludge should be adjusted approximately to a load rate under which the prototype is expected to operate.

Boyle and Rohlich [203] analysed the scaling problems related to activated-sludge treatment systems starting from experience gained in the fermentation industry. They suggested laboratory experiments to study the effects of nutrient concentration and composition on treatment efficiency, underlining the importance of proper sludge adjustment at the laboratory, in order to obtain a biomass corresponding to prototype conditions. The laboratory experiments should therefore be designed to produce approximately identical ambient conditions (pH, temperature, nutrient and dissolved O_2 concentration, etc.) in the model and prototype alike. A special merit of the paper is the appraisal of the different reactor types, such as batch, semi-continuous and fully continuous operation, and the applicable scaling relations. The book by Johnstone and Thring is adopted as the basis in the study of scaling batch-type reactors for the purpose of designing systems with continuous operation. They pointed to the catalytic effect of the reactor wall from the viewpoint of reaction kinetics. In laboratory reactors the surface: volume ratio A/V is normally higher than in prototype units. In geometrically similar systems the basin surface per unit fluid volume decreases according to the -1st power of the characteristic dimension L, so that $\lambda_{A/V} = \lambda^{-1}$. For this reason Johnstone and Thring suggested inactivating the reactor wall. Moreover, they specified approximately identical sludge ages in the model and prototype units. To comply with this specification it is necessary, however, to also reproduce the retention time distribution of the flowing medium with due regard to the prevailing hydraulic conditions, such as stagnation, areas of reduced flow velocity, etc. The tests to be performed during the operational trials of the treatment plant are finally stated to be important tools of the designer to check the results of laboratory experiments and the scaling method adopted.

Eckenfelder *et al.* [204] in 1972 presented a comprehensive review of the concept developed for scaling biological wastewater treatment reactors. They concentrated attention on three main aspects, namely

(a) the effect of bacterial population;
(b) the role of turbulence and
(c) the nature of the departure from the ideal reactors.

Concerning (a) the realization of the "identical biomass" which is theoretically possible irrespective of scale is emphasized. The authors conceded the difficulty of taking the effect under (b) into account in the scaling process. By the simultaneous evaluation of sludge settling and oxygen transfer it is theoretically possible to determine the optimum turbulence level which is fairly simple to adjust in the model. In this connection three methods are mentioned, namely the descriptions by the *Re* number, the *N/V* ratio and the extended mass transfer coefficient $K_L a$. In support of the applicability of the method mentioned last, Eckenfelder *et al.* referred to the paper by Bartholomew. The role of the retention time and time distribution (stagnation areas, short-circuiting, etc.) is finally mentioned under (c) without allowance for which technological scaling is conceivable. At the same time it should be pointed out that, as demonstrated by experience, apparently successful scaling of treatment units does not always result in identical micro-organism behaviour in the model and prototype systems, the behaviour being controlled by the initial and boundary conditions of the unit operations among others.

Eckenfelder *et al.* considered laboratory and/or pilot-scale experiments necessary mainly for solving the following problems:

(a) comparison of alternative technologies of treatment;
(b) determination of design parameters and coefficient, such as kinetic

constants, the necessary rates of oxygen and power supply, the sludge volume produced;

(c) identification of any operational difficulty, such as toxic effects and temperature, foaming, etc.

Sporadic references can also be found in the literature on the subject to the comparison of laboratory and prototype systems. Mancini and Barnhart [205] compared the results of laboratory and prototype measurements on aerated ponds constructed for the treatment of industrial wastewater. They found good agreement between the parameters of reaction kinetics. Quirk [206] compared pilot aeration tank volume = 2×55 gal and full scale activated sludge systems by determining the relationship between the retention time and the removal efficiency.

Gates et al. [207] studied the kinetics of bacterial processes under laboratory and prototype conditions. The laboratory experiments were eminently suited to substantiating the applicability of Monod's kinetic model. The results were checked by measurement in the prototype. Although these studies were concerned primarily with stream simulation the methods adopted are of interest from the viewpoint of biological wastewater treatment. A remarkable feature of the paper is that measurement data are given on the relationship between the intensity of turbulence and bacterial activity, describing the turbulence intensity in terms of the overall transfer coefficient of oxygen $K_L a$.

Attention is finally given to a few publications describing experimental equipment for laboratory and pilot-scale studies on biological wastewater treatment [130, 208, 209].

A description of liquid feed (water, sewage, chemicals, etc.) and flow rate control model equipment was published by research staff members of the Water Pollution Research Laboratory [210].

Mathematical modelling of activated-sludge systems

The question of mathematical modelling and simulation of wastewater treatment units is but loosely connected to the subject under consideration. For this reason attention will be directed only to a few recent publications which contain information of potential interest in similarity and scaling problems.

Naito et al. [211] examined mathematical modelling as a tool for studying biochemical oxidation by taking experimental data into account. Burkhead and Wood [212] developed a computerized method for the analogue simulation of activated sludge systems under steady and unsteady operating conditions. Theoretical and practical examples of modelling mathematically the operation of biological treatment plants (including the settling tanks as well) were presented by Silveston [213]. In their detailed study Erickson *et al.* [214] approached modelling and optimization by systems analysis. The applicability of mathematical, kinetic models was checked by Tebbutt and Christoulas [215] underlining the importance of laboratory and pilot-scale equipment, which can be applied to advantage instead of prototype units, in problems of practical, scientific and even economic nature. Weston *et al.* [216, 217] devoted several papers to the connections between batch and continuous-treatment units, mainly adopting a kinetic approach to the use of laboratory data in designing prototype units. The comprehensive study of Blanch and Dunn [218] is finally mentioned in which examples are given for the solution by mathematical modelling and simulation of biochemical engineering problems, including the case of activated sludge reactors.

2.3.3 Rotating-disc biological equipment

Publications on biological rotating discs used in wastewater treatment have only appeared in the literature during the past few years. Of these the papers by Chesner and Molof [219] and by Ouano [220, 221] will be mentioned first. The scientific antecedents include among others a few papers by Levics [187] who derived in describing mass transfer processes and diffusion currents dimensionless relations in terms of the Reynolds number for the case of revolving discs submerged in fluids. The result obtained consisted of a relationship between the Nusselt, Prandtl and Reynolds numbers.

In designing practice the peripheral velocity is normally adopted as one of the most important criteria. In their paper referred to before Chesner and Molof attributed inferior importance to the peripheral velocity in scaling problems. An important contribution to the field of modelling was made by Ouano, according to whom the oxygen intake and mass transfer processes taking place in biological rotating disc treatment plants can be scaled with a fair degree of approximation. He adopted Reynolds scaling as the main similitude criterion. As experimental verification he derived the dimensionless combination reproduced in Fig. 37, which is based on the empirical relationship of the form of a power product

$$\frac{K_L V_t / A_t}{D_L} = K \left(\frac{A_t}{A_p}\right) \left(\frac{D\omega\varrho}{\eta}\right)^b \tag{250}$$

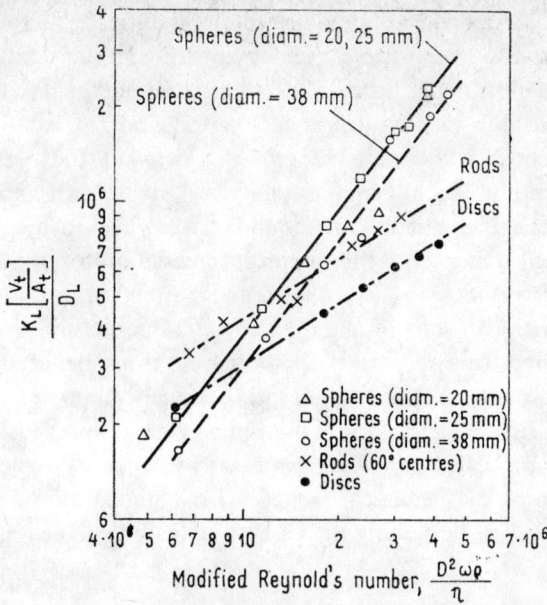

Fig. 37. Effect of *Re* number on the mass transfer coefficient

written on the strength of dimensional considerations and substantially representing the relation of the Nusselt number related to component transfer and the Reynolds number.

The symbols used in Ouano's original paper are as follows:

K_L = liquid film coefficient;
V_t = net tank volume;
A_t = surface area of the tank;
A_p = area of the set of discs projected to the tank surface;
D = disc diameter;
D_L = diffusion coefficient;
ω = angular velocity of rotation;
ϱ = density of the fluid;
η = dynamic viscosity of the fluid.

The power b can be evaluated from Fig. 37. (It should be noted that several alternatives are shown in the figure, namely disc and rod systems and a design formed of spheres, for a detailed description of which reference is made to the paper of Ouano.) For rotating disc equipment the exponent $b = 0.59$ is obtained for the case A_t/A_p = constant (which is logical to assume in a system of given geometry). Note that the experiments of Bintanja *et al.* [222] have yielded a square-root expression relating K_L and the speed of rotation, i.e. $b=0.5$.

From now on for the method adopted for the similitude analysis of the oxygen transfer processes in aeration tanks, the relationship between the similarity transformation parameters (scale factors) for the case of biological rotating disc treatment units will be considered. Assuming Eq. (250) represents an invariant function, for geometrically similar systems one obtains

$$\lambda_D = \lambda \qquad \lambda_{A_t} = \lambda_{A_p} = \lambda^2 \qquad \lambda_{V_t} = \lambda^3.$$

Assuming the same flowing medium in the model and prototype gives

$$\lambda_{D_L} = \lambda_\varrho = \lambda_\eta = 1$$

and thus the scale factors are related as

$$\lambda_{K_L} = \lambda^{2b-1} \lambda_\omega^b. \tag{251}$$

With the assumption $b=0.5$ (Bintanja *et al.*)

$$\lambda_{K_L} = \lambda_\omega^{1/2} = \lambda_t^{-1/2} \tag{252}$$

which is in complete agreement with Eq. (189) derived from Danckwert's Eq. (188).

2.4 SLUDGE TREATMENT

Of the unit operations involved in wastewater and sludge treatment the scaling problems of sludge treatment have received least attention. Two technological steps, however, anaerobic sludge treatment and sludge thickening, are considered important enough to be considered here.

2.4.1 Digesters

In digester scaling problems the following processes command greater attention: agitation (stirring) of non-Newtonian media, mass and heat transfer, reaction kinetics (and the related phenomena such as gas evolution). At the moment the information available in the literature on the subject, especially on the behaviour of structure-viscous media, is too scarce to permit any reliable treatment. Nevertheless, it is advisable to establish some of the basic principles which may serve as starting points in further research. In this respect the experience from the fermentation industry may offer welcome guidance [223].

Reynolds numbers in non-Newtonian fluids

Investigations into the handling and agitation of non-Newtonian fluids have warranted the introduction of several modified forms of the *Re* number, which may be regarded as potential foundations of scale-up criteria. For plastic and pseudoplastic (structure-viscous) materials the modified or generalized *Re* numbers have been derived by considering the models suggested by Bingham and Ostwald–de Waele (see Fejes [290]). The *Re* number written in terms of the shear velocity $U_* = \sqrt{(\tau_0/\varrho)}$ according to Bingham's model is

$$Re_\tau = \frac{U_* l}{\nu} = \frac{\sqrt{(\tau_0/\varrho)}\, l}{\nu} \tag{253a}$$

where l is the characteristic length; and τ_0, the flow limit, i.e. the lowest shear velocity needed to induce flow (for $\tau < \tau_0$ the flow velocity is zero).

In some publications the square of the preceding Re_τ number, denoted as Re_τ^* is found as the characteristic quantity

$$Re_\tau^* = Re_\tau^2 = \frac{\tau_0 l^2 \varrho}{\eta^2}. \tag{253b}$$

Starting from the Ostwald–de Waele material model Metzner *et al.* (see Fejes [290]) have written the generalized Re_n number in the form

$$Re_n = \frac{v^{2-N} l^N}{K/\varrho} \propto \frac{n^{2-N} d^2}{K/\varrho} \tag{253c}$$

where N is the coefficient of structural viscosity; and K, the rigidity coefficien (stiffness factor).

The right-hand side of the proportionality contains the Re_n number applying to the operation of agitators. In Newtonian fluids $N=1$ and $K=\eta$.

Scaling on the basis of the modified Re numbers

Among the scaling criteria related to the agitation of non-Newtonian fluids, such as some waste sludges, invariance of the modified *Re* number plays an eminent role. Assuming the validity of the Ostwald–de Waele relationship, from Eq. (253c) one obtains the scale factor of agitation speeds and velocities. If media having identical properties are used in the model and prototype ($\lambda_\varrho = \lambda_K = \lambda_N = 1$)

$$\lambda_v = \lambda^{-N/(2-N)} \quad \lambda_n = \lambda_d^{-2/(2-N)}. \tag{254a,b}$$

From this it is inferred that although materials having identical properties have been assumed in the model and prototype, the magnitudes of the scale factors related to the operating variables are influenced by the coefficient N of structural viscosity.

It should be noted that by adopting an approach similar to the above the scale factors of other variables can also be derived from the invariants (253a–c).

Erdmenger's concept

Special importance is attributed to the achievement of Erdmenger [224] on the agitation of highly viscous (non-Newtonian) media. In his paper published in 1964 he succeeded in satisfying simultaneously, with fair approximation, the following two relations in the model and prototype (with single prime denoting the prototype and double prime the model quantities):

for the unit technological output

$$\left(\frac{G}{V}\right)' \approx \left(\frac{G}{V}\right)'' \tag{255a}$$

for the unit electric power demand

$$\left(\frac{N}{V}\right)' \approx \left(\frac{N}{V}\right)'' \tag{255b}$$

where G is the technological output of the continuous agitator (kg/h); V, the net volume of the agitator (m^3); and N, the electric power (kW).

In view of the fact that the unit work expended for agitation, expressed in kWh/kg, may also be interpreted as the ratio of Eqs (255a) and (255b), it was established that if the geometrical similarity requirement is satisfied and identical kinds of fluid are used in the two systems, the unit value of agitation work was constant in the model and the prototype.

Under specific conditions simultaneous validity of Eqs (255a) and (255b) can be demonstrated by a theoretical approach as well. Assuming, for example, that the flow of the highly viscous medium agitated is a laminar one and that the two systems are geometrically similar, the electric power demand N can be shown to be proportionate to the volume V

$$N \propto n^2 d^3 \propto n^2 V \tag{256}$$

where n is the speed (rpm); and d, the diameter of the stirrer (m).

Assume further that the scale factor of time t is unity (thus $\lambda_t = 1$ and $t' = t''$), so that the scale factor of speeds is also unity ($\lambda_n = 1$). Under these conditions the outputs G related to unit volume will also be seen to be identical ($\lambda_G = \lambda_V$). On the other hand, combining Eqs (255b) and (256) it can be immediately seen that the ratio of electric power demands related to unit volume is also identical in the model and prototype ($\lambda_n = 1$ and thus $\lambda_N = \lambda_V$).

It seems necessary to emphasize here that these conditions cannot be generalized to all cases. As will be shown later, observation of the requirement $\lambda_t = 1$ is advisable mainly for technological reasons when scaling digesters, but may sometimes produce entirely misleading results.

Implications in the fermentation industry

Concerning industrial fermentation processes the book by Aiba *et al.* [156] is referred to first. The chapter on scaling presents a brief discussion of the modelling problems associated with the treatment of fermentation media displaying non-Newtonian behaviour. The authors quoted several practical examples where scaling problems were solved successfully on the basis of equal unit output and the extended mass (oxygen) transfer coefficient. It emphasizes that continued research is needed to gain better understanding of the scale-up problems arising in fermentation processes of non-Newtonian media.

In this context only a brief mention is made of the paper by Deindoerfer and West [223], who studied the rheological properties of fermentation media. The comprehensive review of the dimensionless numbers characterizing Newtonian and non-Newtonian (plastic and pseudo-plastic) media is believed to be of special interest as it may form the starting basis of model studies into the processes under consideration.

Implications in anaerobic sludge treatment

The present state of scaling the units of anaerobic sludge treatment has been reviewed briefly in a paper by Jones [130]. In research practice laboratory-scale equipment is normally used, e.g. for studies on reaction kinetics (including the process of gas production) or for predicting the effects of different operating conditions. This is exemplified by the paper of Ward [225]. An interesting comparison of the results of laboratory, pilot and plant-scale studies has been presented by Bayley [226]. Concerning the comparison of pilot and full-scale data, the studies of Torpey [227] are of note, who besides providing remarkably valuable information on the interrelations between digesters of different size, described classical examples of the misinterpretations made in scaling technological processes and equipment without regard to similitude considerations. This is clearly reflected by the paper referred to and the comments made on it. The main conclusions will be summarized subsequently.

Torpey in his study examined the allowable load on the digester system at one of the sewage treatment plants of New York City. In these studies a pilot unit was also used parallel to the prototype, in order to examine the allowable load in a wider range. He introduced the concept of the Digestion Index as

$$DI = \frac{GQ}{gV} \times 100 \quad (\% \text{ per day}) \qquad (257)$$

where G is the volume of gas produced daily by the digester; V, the rated volume of the digester; and g, the quantity of gas present in the sludge that leaves the digester.

Torpey expressed the efficiency of the digester in terms of the Digestion Index. Measurements have revealed differences in the value of the Digestion Index in the pilot and prototype units, although the retention times and unit loading rates were identical. The differences persisted regardless of the efforts to maintain within practicable limits identical ambient conditions (temperature, pH, etc.) in the different scales. Torpey attributed the deviation in the numerical value of the Digestion Index primarily to the difficulty of achieving in the prototype a stirring intensity sufficiently high to use effectively the entire digester volume. This implied that stagnation areas (e.g. bottom deposits) of appreciable magnitude developed in the prototype unit.

In his comments on the preceding study Schlenz [228] questioned the validity of some of Torpey's conclusions on grounds pertaining essentially to the domain of scaling. He pointed out that the ratio Q/V involved in the expression defining the Digestion Index is constant, provided that—as in this particular case—the average retention times are identical in the pilot and prototype units. Consequently, the deviations in the value of the Digestion Index stem from variations in the ratio G/g. In another contribution Heukelekian [229] clearly formulated the question of whether it is at all possible to transfer the data obtained in a pilot unit to prototype conditions. No definite answer was given to this question during the discussion. Nevertheless, these references are highly instructive by illustrating the deficiencies which may result if scale-up is performed without observing similitude considerations. It should be clearly understood that the value of some variable characterizing operation, such as the Digestion Index, is not necessarily identical in different scales (even if the scale factor of the average retention time is unity).

This deviation may be traced back to several causes. The first of these is that according to the theorems of similitude, the scale factor of some variables may be unity, while that of others may depart from this. The other potential cause may be gross infringement on geometrical similarity. Moreover, failure to realize similar ambient and operating conditions (identical temperatures, stability of loading, etc.) in the pilot and prototype units has also been observed frequently in practice.

Mention should also be made of the paper published by Dick and Ewing [230] in 1967, who analysed the rheological properties of activated sludge, founded on both theoretical and experimental investigations. These aspects should also be included in scaling digester processes.

Experience has shown the difficulties encountered in scaling digesters to stem from the following causes: the material properties of the sludges to be treated (structure-viscous media) are inadequately understood in general, and even these may vary in time. There is a lack of equations describing the stirring, mass and heat transfer processes in large sludge volumes. It will be readily perceived that in digesters of several thousand cubic metres volume, stirring conditions resembling those in fermenters of a few litres size are impossible to realize. From the above it can be concluded that the present scaling methods are purely empirical in character and are not founded on similarity theory.

Scaling based on similitude considerations and empirical data

An attempt will be made at developing a scaling method which is applicable to problems encountered in practice and is founded on similitude considerations, as well as on data determined empirically.

As will be recalled, when using the conventional modelling relationships, such as the Froude or Reynolds, etc., similarities, conversions between the model and prototype are made by using the scale factors. Practically this means that the scale factor $\lambda_x = x'/x''$ of the variable x is related to the scale factor of lengths, for example. Relations of this kind may be derived by assuming the validity of a particular model law. To illustrate, when observing Froude similarity in the familiar way, the scale factor λ_v of velocities equals the square root of the scale factor λ of lengths. If it is desired to check whether a particular model law is valid or not in a particular case, a potential avenue consists of performing measurements in units constructed to different scales and finding the power of the scale factor λ empirically from the measurement data. This approach offers the possibility of comparing the calculated value of the exponent with theoretical values derived on the basis of different model laws, further of selecting the one resulting in closest agreement between the theoretical and empirical values (see also Section 2.2.6). Evidently the same philosophy can be applied to distorted models as well.

In analysing the scaling criteria of digesters it is convenient to assume that the digester units applied in practice are in some way distorted replicas of each other. This assumption is all the more justified, since as will be shown the data available for processing apply to geometrically dissimilar systems. At the same time it is necessary to emphasize here that scaling problems could be solved with much less effort and more accurately if uniform data series on geometrically similar systems would be available for evaluation.

Using the above arguments, the following approach will be adopted. The scale factor λ_x of a variable x will be related to the scale factor $\lambda_V = V'/V''$ of volumes. As will be shown later, one may write the relation in question in the form of a power function

$$\lambda_x = \lambda_V^{\alpha_x} \tag{258}$$

where the power α_x plays a dominant role and is evaluated from empirical data conveniently by a graphical method.

Obviously when considering geometrically similar units, the familiar relationship

$$\lambda_V = \lambda^3 \tag{259}$$

applies and can be used in combination with Eq. (258).

The theoretical solution outlined cannot be applied in practice unless the powers α_x are known. To evaluate these, published measurement data have been processed and analysed. Of the publications used the paper by Wiedemann [231] is mentioned first, but interesting series of data have also been obtained from Pallasch and Triebel [232] on the performance of MAN-type sludge stirring equipment. Measurement data obtained by myself at sludge treatment plants in Hungary have also been included in the analysis [233].

Over 300 results collected have first been grouped according to the main types of digester design. In this way four groups have been distinguished
  i. Digesters with axial-flow pump agitators.
  ii. Digesters with sludge circulation by means of air-lift pumps and with external sludge pumps.

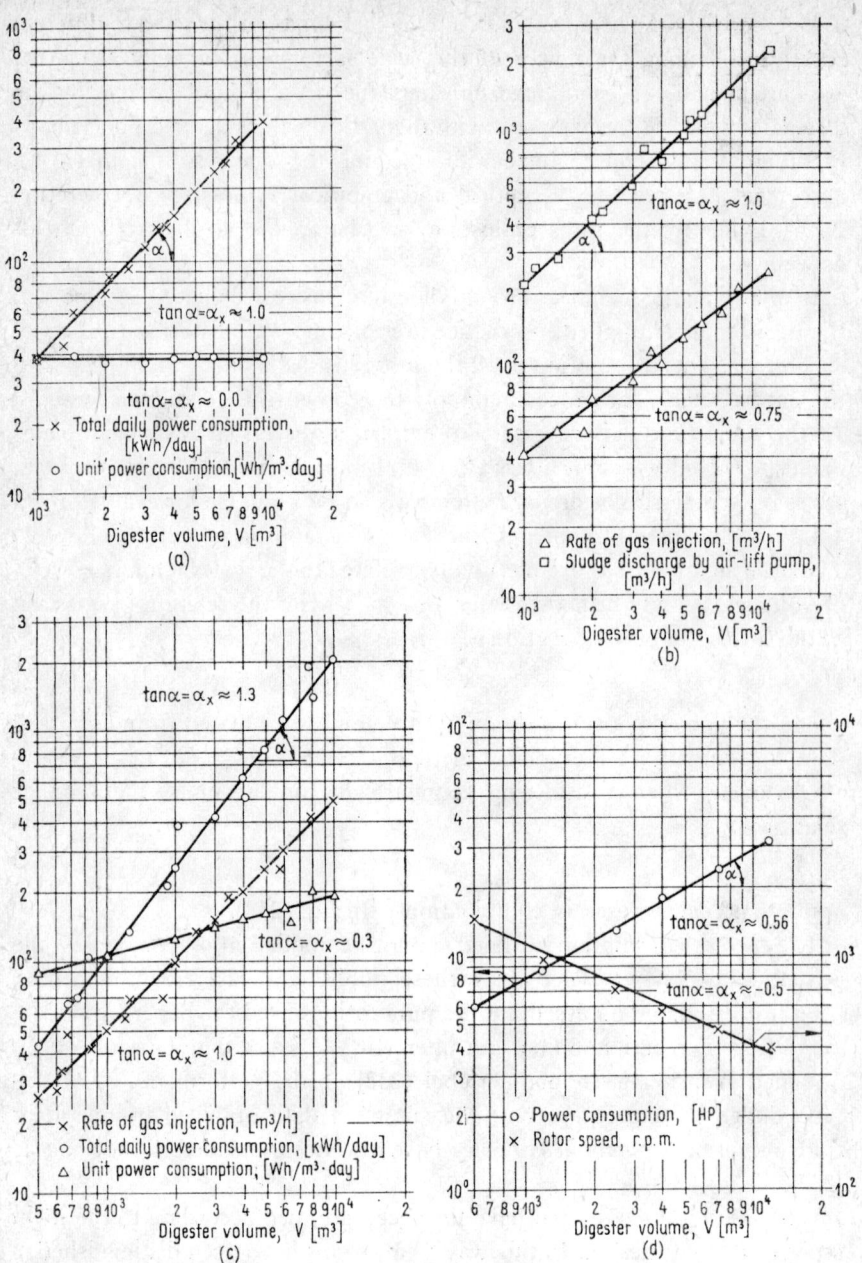

Figs 38(a)–(d). Mechanical stirring with fixed-blade pump. Agitation with air-lift pump and external sludge pump. Agitation by gas injection and external sludge pump. MAN-type sludge stirring rotor

iii. Digesters with gas injection and external sludge pump.
iv. Digesters equipped with MAN-type sludge circulation rotors.

The main parameters of the above groups with volumes from 500 to 10,000 m³ are as follows:

i. The amount of sludge corresponding to the digester volume is recirculated six times during 24 h. Stirring is intermittent for periods varying from 3 to 10 h daily. The axial-flow pump is mounted in a vertical pipe inducing upward flow therein.
ii. A vertical pipe is again installed in the system for sludge circulation by the air-lift pump. The flow pattern is, however, a complex one, produced by the circulating effect of the external sludge pump. The sludge volume is recirculated round six times during 24 h. The air-lift pump operates almost continuously.
iii. In this group the gas is introduced at several points distributed over the bottom. No guide pipes are used. The gas is introduced at the rate of about 0.05 m³/m³/h (related to the digester volume).
iv. More detailed information on the design of digesters involving the MAN-type stirring system is to be found in Pallasch and Triebel [232].

The data collected on the systems outlined above have been processed graphically. Without striving for completeness four diagrams are shown in Figs 38a–d together with the data series used. The values of the exponent α_x involved in Eq. (258) have been determined by computing the slopes of the fitted straight lines, with the results compiled in Table 6.

In the course of evaluation the data were found to scatter considerably, but this was expected, since the data series processed were not uniform. In this respect the differences in digester design (different kinds of distortion),

Table 6

Exponents of scaling-up relations for digesters

Relevant variable, x	Exponent α_x
I. Mechanical mixing by axial-flow pump	
Sludge discharge, m³/h	1.0 (1.25)
Total daily power consumption, kWh/day	1.0
Unit power consumption, Wh/m³ day	0.0

Relevant variable, x	Exponent α_x
II. Mixing by air-lift pump combined with external sludge pump	
Discharge of extl. sludge pump, m³/h	1.0 (1.3)
Discharge of air-lift pump, m³/h	1.0
Rate of gas injection, m³/h	0.75
Gas pressure, bars	0.38 (0.37–0.44)
Total daily power consumption, kWh/day	1.0
Unit power consumption, Wh/m³ day	0.0
III. Mixing by gas injection combined with external sludge pump	
Discharge of extl. sludge pump, m³/h	1.0 (1.25)
Rate of gas injection, m³/h	1.0
Gas pressure, bars	0.38
Total daily power consumption, kWh/day	1.3
Unit power consumption, Wh/m³ day	0.3
IV. MAN-type sludge mixing rotor	
Sludge discharge, m³/h	1.0 (0.86)
Power demand, kW	0.56
Mixed speed, rev/min	−0.46 (−0.5)

Note: The data in parentheses were also obtained in specific cases. However, their use in computation is not warranted, unless validity limits are examined.

the scale effect and the influence of other relevant variables obviously play an important role. Nevertheless, fitting and the construction of trend lines presented no difficulties.

Conclusions and recommendations on digester scaling

The conclusions and recommendations to be summarized subsequently can be classified broadly into two groups. Some of them are conclusions deduced from the empirical data and already outlined; the second, smaller part comprising those on which no actual measurement data were available and derived by considerations of similitude. These conclusions are presented here in the order of growing complexity in the pattern of similitude theory [233, 234].

(a) *Geometric similarity*

By comparing the results of pilot and full-scale studies it is concluded positively that digesters can be scaled up more accurately if the two systems are geometrically similar. Nevertheless, the chemical, biochemical and specifically the gas production processes can also be reproduced successfully in distorted models.

(b) *Kinematic similarity*

Insofar as the objective is to achieve technological similarity, the criterion $\lambda_t = 1$ is of paramount importance. In practice this implies that the scale factor of the characteristic times, such as retention times, in the corresponding model and prototype systems should be unity. The situation is, however, a different one if, for instance, the process or the intensity of stirring is considered. In such cases the scale factor of the characteristic times can be derived from the criterion of dynamic similarity to be discussed later, for instance by introducing the modified Reynolds number already mentioned. In connection with kinematic similarity it appears logical to extend considerations to the possibilities of converting other characteristic quantities, the units of measurement, or dimensions which are some power function of the dimension of time t. In this respect the speed of the sludge stirring equipment is mentioned first. The scale factor of speeds has been found in practice to be determined expediently by experiments. The fundamental objective here is to apply speeds in the model and prototype, which ensure at least with fair approximation circulation of the same intensity in the entire digester volume. This statement also applies logically to gas-injection and combined systems.

(c) *Dynamic similarity*

Adherence to the criterion of dynamic similarity is essential mainly for reproducing hydraulic phenomena. In scaling digesters, the solution of the problem is greatly complicated by the fact that the medium is a non-Newtonian one, the physical, material properties of which are little, if at all, understood. The measurement data appear to support the earlier conclusion, according to which in scaling digester systems the modified Reynolds number, as the dominant invariant, may be adopted as the basis of the scaling relationship. The same is implied also by the paper of Deindoerfer and West referred to earlier [223]. It should be noted, however, that further research is necessary to determine the criteria of dynamic similarity more accurately.

(d) *Physico-chemical and biological similarity*

Virtually no experimental data are available on this subject, although model and pilot scale investigations are widely used. Of the physico-chemical processes, the problems related to scaling up mass transfer phenomena may be mentioned first. The logical approach to get physico-chemical and biological similarity appears to involve the method of trivial modelling. In this particular case it implies that the same physico-chemical, chemical and biological parameters must be ensured in unit volumes of the model and prototype systems in order that the processes of anaerobic fermentation should be comparable in the two. Formulated in a different way, the method of trivial modelling is applied correctly if the medium in the model (and the processes taking place therein) may be regarded a specific part volume of the prototype system. However, this requirement cannot be adhered to, unless additional criteria are also satisfied. In this respect the importance of stirring at the same intensity is emphasized.

(e) *Technological similarity*

In accordance with the principles of similarity theory outlined above, the similarity criteria related to the individual component phenomena of the technological process must take account of those predominating in a particular problem. The set of similarity criteria characterizing the individual unit operations should then be examined in establishing the scaling and similitude equations of the treatment method. This line of reasoning has prompted the introduction of the concept of technological similarity. By processing the empirical data available it could be established rather positively that the scale factor of sludge yields (sludge outputs, m³/h) corresponds to the scale factor of volumes V, since in all of the four systems studied the exponent α_x was found to equal unity. The essential conclusion arrived at was that the unit load Q_i/V related to unit digester volume must be invariant (have the same value in the model and prototype) and thus the requirement $\lambda_i = 1$ can also be satisfied. Examination of a number of actual measurement data suggested that adherence to this invariance may be one of the fundamental requirements in the technological scaling of digesters.

(f) *Economic similarity*

In a paper I published recently I introduced the concept of economic similarity, illustrating the application of the similarity principle to scaling-up the dimensions and capacity of various sewage treatment systems by actual examples [235]. Starting from the reasoning outlined there and in accordance with the considerations summarized in Chapter 4, one may deduce conclusions of interest in this subject. In the case of alternatives i, ii and iii the total daily and the unit power consumption have been adopted as the variables extrapolated on the basis of similarity criteria. As will be seen from Table 6 the exponent α_x assumes the value 1.0 for the total daily power consumption in alternatives i and ii. In alternative iii the same exponent was 1.3. On the other hand, the exponents for unit power consumption were 0.0 and 0.3. These data are in fair agreement with each other, in so far as the difference of the corresponding exponents of total daily and unit power consumption is unity. As will be recalled, in cases where the exponent α_x related to cost-proportionate daily power consumption is unity, the costs are termed proportionate. If this value is greater than unity, such as 1.3 in alternative iii, the costs are termed progressive. The case $\alpha_x = 1$ represents the alternative, including also the criterion based on the invariance of unit sludge load that has been mentioned when commenting on the scientific results of Erdmenger.

2.4.2 Thickeners

The comparison of laboratory, pilot and prototype studies on the thickening process in secondary settling structures at activated-sludge treatment plants has been scrutinized in the discussion on the paper by Hibberd [236]. Attention has been concentrated on the vertical stratification of the sludge, emphasizing the importance of using experimental equipment in which the process is accessible to visual observation (models with transparent walls).

The modelling techniques and the potential scaling methods of gravitational sludge thickeners have been reviewed and perfected by Edde [237] and Edde and Eckenfelder [238]. Edde's doctoral thesis introduced new aspects to the literature on thickeners.

Stalmann [239] examined the dimensions (depth, volume) of thickeners for their role in, and effect on, the technological processes, though mainly from viewpoints unrelated to scaling.

A laboratory method of scaling thickeners is finally described after Eckenfelder and Ford [240], which can also be used to advantage in experiments to establish design criteria.

The main variables characterizing the thickening process are related by the empirical expression

$$\frac{C_u}{C_0} - 1 = \frac{K_b}{ML^n} \qquad (260)$$

where C_0 is the initial concentration of sludge entering the thickener; C, the concentration of the (thickened) sludge removed from the thickener;

ML (Mass Loading), the surface suspended-substance loading (kg/m³ per day); and n, K_b, experimental constants.

The size of n has been shown to depend primarily on the rheological properties of the sludge and to be largely unaffected by the dimensions of the experimental unit. The value of K_b is influenced, on the other hand, by the very same variables.

On the strength of these considerations the scaling factor K_T/K_b has been introduced, where K_T refers to the prototype conditions while K_b to the experimental unit, as considered earlier. From investigations on thickeners the factor K_T/K_b has been related empirically to sludge depth, which made numerical evaluation possible of the scale-up factor pertaining to particular sludge depths.

As an illustrative example Figs 39a and b are given (after Eckenfelder). From (a) the value of K_b is found for known C_0 values (on the basis of

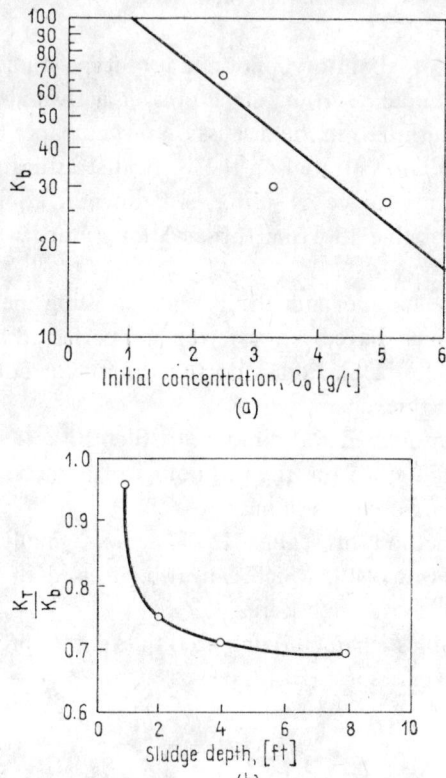

Figs 39(a) and *(b)*. K_b vs the initial solids concentration. Scale-up factor K_T/K_b vs sludge depth in the thickener

laboratory experiments), while (b) yields the factor K_T/K_b for sludge depths specified in advance. Multiplying the scaling factor by K_b, one directly obtains K_T, which when substituted (in lieu of K_b) into Eq. (260) yields the surface (suspended-substance) mass-loading rate ML. (Evidently the exponent n must be found by experiments.) For a given loading rate the area of the thickener can finally be estimated.

3. Applications of similitude in activated sludge treatment

In this chapter details will be given on the research work carried out by theoretical and experimental approaches during the past 15–20 years into the problems of modelling and scaling encountered in activated sludge wastewater treatment. No mention will be made here of the related development work, as a result of which, founded on similarity theory, new products have been designed [241, 242], since this would exceed the scope of this book. References to the patents registered in the course of theoretical modelling work are nevertheless warranted, since it is firmly believed that realistic harmony between theory and the requirements of practice is especially desirable in this field.

3.1 HYDRAULIC PROCESSES

Studies into the hydraulics of aeration basins and tanks of different type and geometrical design have provided opportunity to determine the validity of the model laws applying to the movement of various media. Statistically processed data series obtained by measurements in two or three geometrically similar units of different size have been used for this purpose. By measuring the velocities in discrete points of the flow field, and by determining the velocity ratios at corresponding points of the models and prototypes of different size, it is possible to calculate the scale factor λ_v and thus to check for its validity the model relationship assumed. The theoretical justification of this approach is the fact that the velocity ratio can be written as the power of the scale, specifically the scale factor λ of lengths ($\lambda_v = v'/v'' = \lambda^{\alpha_v}$).

For some of the aeration reactors more widely used in wastewater treatment practice the following values have been obtained for the exponent α_v:

INKA-type air injection basin [243, 244] (see Fig. 40) $\alpha_v \approx 0.43$
Kessener-type basin [245] $\alpha_v = 0.475$

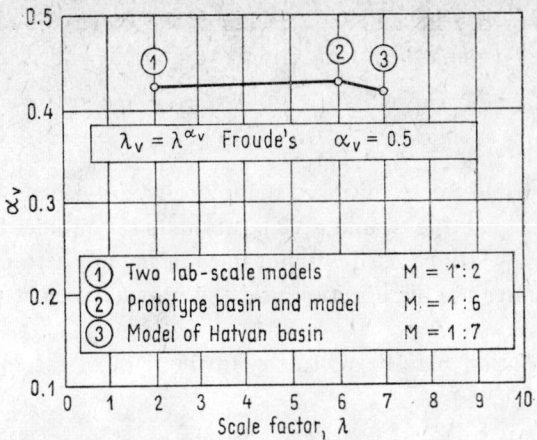

Fig. 40. Scale factor λ vs α_v in INKA-type aerators

Aeration tank with "Mammoth-rotor" [246] $\alpha_v = 0.56$
Aeration tank with vertical-shaft rotor [182, 183] $\alpha_v = 0.58$

It can be seen that hydraulic conditions can be reproduced with a fair degree of approximation on the basis of the Froude similarity. The theoretical value in this case would be $\alpha_v = 0.5$. The discrepancy between the theoretical value and those actually observed implies—in agreement with other authors—that besides gravitational and inertial forces, other effects are also present.

It should be noted that in the above experiments the operating variables have been adjusted *a priori* on the basis of the invariance of the *Fr* number, thus in the case of aerators with horizontal and vertical shaft arrangements $\lambda_n = \lambda^{-1/2}$. In the case of the INKA-type basin certain corrections proved necessary to account for the movement of bubbles. The principles may be summarized as follows [183]:

As already mentioned in Section 2.3.2, in the compressed-air aeration basins of practical interest the greater part of the air volume introduced is present as bubbles having diameters larger than 2.5 mm. Consequently it is deemed permissible to assume the same rising velocities for the bubbles in the model and prototype alike, irrespective of the fact that the bubbles moving in the model are smaller than those in the prototype. Accordingly

$$w' = w''.$$

With regard to this it is possible to determine approximately the conversion factor of the rate of air supply injected into the basin. In view of the fact that the bubbles rise in a flowing medium, the mean velocity v_m of the medium (water) passing the air injection grid must be added to the velocity w, thus

$$v_B = w + v_m. \tag{261a}$$

Evidently it would be a mistake, theoretically, to neglect the vectorial character of the quantities w and v_m. The absolute magnitudes of the preceding velocities may still be used without introducing any serious error, since immediately above the air injection grid the velocity vectors w and v_m are virtually vertical.

The same relation may be written with the data of the model and the prototype

$$v'_B = w + v'_m \quad v''_B = w + v''_m$$

so that

$$\lambda_{v_B} = \frac{w + v'_m}{w + v''_m}. \tag{261b}$$

Assuming Froude similarity to apply to the flow of the liquid medium, one has $\lambda_{v_m} = \lambda^{1/2}$. Upon substitution the ratio of the velocities of bubbles moving in the flowing medium becomes

$$\lambda_{v_B} = \frac{w + v''_m \lambda^{1/2}}{w + v''_m} = \frac{w + v'_m}{w + v_m \lambda^{-1/2}}. \tag{262}$$

The scale factor of air supply rates is therefore found from the expression [183]

$$\lambda_{Q_{\text{air}}} = \lambda^2 \frac{w + v''_m \lambda^{1/2}}{w + v''_m} = \lambda^2 \frac{w + v'_m}{w + v'_m \lambda^{1/2}}. \tag{263}$$

As regards the velocity v_m involved in the preceding expressions it should be noted that it is rather difficult to determine in the prototype while fairly simple in the model. Allowance may further be made for the fact that any error of measurement in the magnitude of v_m entails no serious error in the value of $\lambda_{Q_{\text{air}}}$ within the range $\lambda < 10$, where model studies on aeration basins are normally performed. For example, in the case of $\lambda = 2$ the effect of v''_m on the magnitude of $\lambda_{Q_{\text{air}}}$ will readily be seen to be insignificant, while in the case of $\lambda = 10$ this effect is already an appreciable one. For instance, if v''_m is substituted with 35 instead of 30 cm/s, the resulting error in $\lambda_{Q_{\text{air}}}$ will be round 5% and acceptable. At λ values smaller than 10 the error will be even smaller.

Equation (263) is obviously related to the Froude similarity requirements, since in writing Eq. (262) the validity of the Froude similarity criteria has been assumed. At the same time, Eq. (263) also contains the correction for the rising bubbles. In the course computations the initial acceleration of the bubbles has been neglected and uniform movement assumed. This approximation is justified since the accelerating section of movement is terminated over a pathlength of a few hundredths of a millimetre within a few hundredths of a second.

Attention is finally called to the fact that the movement of bubble groups differs from that of discrete bubbles, major differences being observable especially in the case of larger bubbles. The rising bubbles interfere with each other, the larger ones may even overtake and incorporate smaller bubbles. Surface fluctuations in the aeration basin are due mainly to the bursting of such larger bubbles. In a certain range of large bubbles the Froude criterion applies [see Eq. (242)], the process resembling flow in open structures.

3.2 ON MODELLING TURBULENCE CONDITIONS

In the course of model studies on aeration basins the problem of scaling turbulence conditions arises. Any detailed theoretical and experimental investigations into the turbulence conditions in aeration basins would logically cover a very wide field and form the subject of an additional paper. In this instance attention will be confined to the application of a single scaling method only, by which the extent of pulsation can be converted approximately using measurement data in two geometrically similar units of different size.

It may be inferred from the literature on the subject that no solution has been found so far for the exact modelling of turbulent flow phenomena. This is due in part to theoretical difficulties and in part to the problems of measurement. By introducing simplifying assumptions, characteristic dimensionless groups are suggested in diverse publications. Authors normally start from the Reynolds equations and by examining the invariance criteria thereof and by applying similitude transformations derive several dimensionless groups, such as the Strouhal number, the Euler number, the Froude number or the Kármán number. It is of interest to observe that in turbulent flow the Reynolds number, which represents molecular viscosity, is negligible, whereas the Kármán number is of fundamental importance by accounting for the pulsational velocities.

An attempt will be described in the following at representing approximately the turbulent character of flow in hydraulic model tests on vertical-shaft aerators [246, 247]. Two familiar principles will be adopted as the starting basis:

For describing the intensity of pulsation the standard deviation s, or the coefficient of variance c_v calculated from the instantaneous velocities, is considered acceptable.

In establishing the similarity criteria for modelling turbulent flow it is necessary to include, along with the instantaneous values of the variable quantities, also the pertinent stochastic parameters, remembering that a stochastic process is under consideration.

The preceding two principles suggested the use of the coefficient of variance in scaling the intensity of pulsation by similitude transformation. This is justified by the following arguments:

(a) Inclusion of the coefficient of variance in the scaling relations is consistent with the two principles above.

(b) Moreover, the coefficient of variance, as the ratio of the standard deviation and the arithmetic mean, is a dimensionless number and as such may be adopted as a scaling criterion in comparing the processes in systems of different size.

(c) The above statement is supported theoretically by the circumstance that one of the invariance criteria of the Reynolds equations in similitude transformations is derived from the invariant

$$K_{ij}=\frac{v_i^* v_j^*}{\bar{v}_i \bar{v}_j} \quad (i,j=1,2,3) \tag{264}$$

a special case of which is Kármán's number (where $\bar{v}$ is the average velocity, the velocity averaged over time and v^* is the pulsation velocity). In some publications the ratio $K_a = v^*/\bar{v}$ is referred to as the Kármán number, which in the case of $i=j$ is identical with $K_{ii}^{0.5}$. Concerning the expression of the coefficient of variance it is observed that the ratio $v^*/\bar{v}$ is again involved besides the number N of measurement data (the measurement data are velocities in this case).

On the strength of this the coefficient of variance c_v may be regarded a dimensionless number suited to the approximate scaling of pulsation and to allow at the same time for the stochastic character of the process. The resulting scale factors are

$$\lambda_{v^*}=\lambda_s=\lambda_{\bar{v}} \tag{265}$$

in which case
$$\lambda_{c_v}=1 \quad (\text{if } \lambda_N=1). \tag{266}$$

Consequently, in scaling the extent of pulsation the criterion $c_v=$ idem is specified.

Velocities were measured in two geometrically similar basins of different size ($\lambda=2$), the results yielding the values [247, 248]

$$c'_v=4.26\% \qquad c''_v=3.54\% \qquad \lambda_{c_v}=1.2$$

$$s'=3.51 \text{ cm/s} \qquad s''=2.00 \text{ cm/s} \qquad \lambda_s=1.75$$

$$\bar{v}'=82.41 \text{ cm/s} \qquad \bar{v}''=56.51 \text{ cm/s} \qquad \lambda_{\bar{v}}=1.46$$

$$(\alpha_{\bar{v}}=0.54).$$

It is instructive to recall that the investigations described in Section 3.1 yielded $\lambda_v=1.53$ ($\alpha_v=0.58$), while the theoretical value corresponding to Froude similarity is $\lambda_v=1.41$ ($\alpha_v=0.5$). These results imply that

(a) the data obtained from the measurements made in order to characterize turbulence are consistent with the investigations described in Section 3.1, since $\alpha_v=0.58 \approx \alpha_{\bar{v}}=0.54$ and both deviate in the same sense from the theoretical value of 0.5;

(b) one is also free to specify $\lambda_s=\lambda_{\bar{v}}$, in which case $\lambda_{c_v}=1$. The deviation from this condition is 20% in this case. This implies practically that the scale factor of the coefficient of variance c_v determined from the measurement data and adopted to characterize turbulence, differs by 20% from the assumed value, namely unity;

(c) concerning the pulsation velocities characterized by the standard deviation s and the coefficient of variance c_v, approximate similarity of turbulent flows in the model and the prototype units is seen to exist. The measurement data reveal that scaling based on the average velocities is more accurate than that based on the pulsation velocities;

(d) in view of the difficulties in modelling turbulent flow conditions, as well as of the problems of measurement techniques, the approximate scaling method described is believed to be of acceptable accuracy. The method can be extended also to the simultaneous examination of the three pulsation components.

3.3 OXYGEN TRANSFER PHENOMENA

Detailed investigations have been conducted with the aim of describing the oxygen transfer phenomena in different aeration systems. Reports on the results obtained have been presented at international conferences and published in the literature. The main achievements will be summarized below.

From the results of model tests made with clear water, dimensionless groups have been derived, which as invariant functions may serve as the basis of the scaling equations.

(a) For INKA systems [173, 183, 201, 248]

$$\frac{K_L a d_p}{v_B} = 1.048 \times 10^{-2} \left(\frac{h_i}{d_r}\right)^{0.70} \left(\frac{G d_p}{v_B}\right)^{1.12}. \tag{267}$$

(b) For Kessener systems [173, 201, 249]

$$\frac{K_L a}{n} = 1.71 \times 10^{-11} \left(\frac{h_b}{d_b}\right)^{0.94} \left(\frac{n d_b^2}{v}\right)^{1.67}. \tag{268}$$

(c) For vertical-shaft systems [182]

$$\frac{K_L a}{n} = 8.26 \times 10^{-4} \left(\frac{h_b}{d_b}\right)^{0.23} \left(\frac{h}{d_b}\right)^{-0.93} \left(\frac{n^2 d_b}{g}\right)^{0.08} \left(\frac{n d_b^2}{v}\right)^{0.17}. \tag{269}$$

The symbols used are $K_L a$, the extended mass transfer coefficient; d_p, the diameter of ports on the injector grid; h_i, the depth of injector submergence; v_B, the rising velocity of bubbles in flowing media; G, the rate of air supply related to unit basin volume; n, the speed of the aeration rotor; h_b, the submergence depth of the aeration rotor (brush); and d_b, the diameter of the rotor (brush).

Equations (267)–(269) are valid under the experimental conditions described in the previously referred to papers, in the case of clear water flowing in geometrically similar units.

The scaling relationships that can be derived from the invariant functions are

(a) For INKA systems [173, 201] with

$$\lambda_{K_L a} = \lambda_{v_B} = 1 \quad \lambda_{h_r} = \lambda_{d_p} = \lambda$$

Eq. (267) yields

$$\lambda_G = \lambda^{-0.107}.$$

Further the scale factors
$$\lambda_{Q_{air}} = \lambda_G \lambda^3 = \lambda^{2.893}. \qquad (270)$$

For additional experimental checks on the preceding scaling criteria the $K_L a$ values have been measured at different air supply rates in the geometrically similar ($\lambda=6$) prototype unit at the sewage treatment plant of Hatvan town. The scale factor $\lambda_{Q_{air}}$ calculated from the measurement data at five different air supply rates has been compared with the theoretical value of $\lambda^{2.893} = 178.3$ obtained from Eq. (270). The greatest difference between the calculated and measured values was about 9% (see Table 7).

Table 7

Experimental verification of the model law applying to the oxygen transfer conditions in INKA-type aeration tanks

$K_L a' = K_L a''$ (h^{-1})	Q'_{air} (m³/h)	Q''_{air} (m³/h)	$\lambda_{Q_{air}} = \dfrac{Q'_{air}}{Q''_{air}}$ measured	$\lambda_{Q_{air}} = \lambda^{2.893}$ calculated	$\Delta \lambda_{Q_{air}}$ (%)
3.87	6770	38.80	174.5	178.3	+2.13
3.40	6340	37.16	170.6	178.3	+4.32
3.22	6100	35.40	172.5	178.3	+3.25
2.36	4780	26.86	178.3	178.3	+0.17
1.64	3780	19.41	194.6	178.3	−8.98

(b) Kessener system [172, 201]

Starting from the power product of Eq. (268), introducing the scale factors and the criterion $\lambda_v = 1$ the scaling relationship becomes

$$\lambda_{K_L a} = \lambda_{hb}^{0.94} \lambda_{db}^{2.40} \lambda_h^{2.67}. \qquad (271)$$

The applicability of this equation has again been verified in a geometrically similar system of different size ($\lambda=2$).

A plot of Eq. (268) is shown in Fig. 41a. From the measurement data illustrated here the following also may be inferred (beyond the above conclusions).

The experimental data processed in dimensionless form lead to a relationship resembling that of Nikuradse. It is possible to establish the limit value of the characteristic parameters, at which the flow suffers a change in type. These limit values are expressed preferably in terms of the Reynolds number. In this particular case the critical value is $Re = 3 \times 10^5$, at which the inclined straight lines in Fig. 41b start to deflect towards the horizontal.

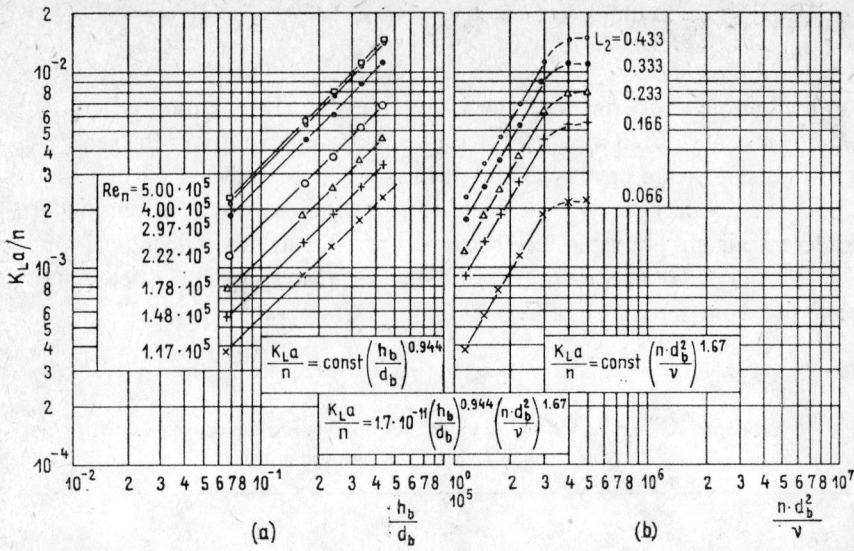

Figs 41(a) and *(b)*. Dimensionless plot of oxygen transfer variables

The curved section implies the presence of a transition range. The $K_L a$ values representing the rate of oxygen transfer are related to the operating variables in different ways below and above the critical Re number. Under turbulent conditions analogous to theoretical hydraulic considerations, the rate of oxygen transfer is unaffected by the Reynolds number.

In order to support the results of experiments made in connection with rotating-brush aerators, some of the more detailed data series published in the literature have also been processed, as shown by the example of the processing work on the measurement results of Baars and Muskat (see [173]). In the manner of Figs 41a and b the dimensionless relationships have been plotted in Figs 42a and b against the $K_L a$ coefficients calculated from the OC values. As can be seen there is a strong resemblance between the families of curves in the two figures. The critical value of the Reynolds number will further be seen to range between the limits 4×10^5 and 5×10^5, depending on the ratio $L_2 = h_r/d_r$, in that the Reynolds number decreases as the submergence depth h_r of the rotor and thus the dimensionless number L_2 is increased. Concerning the $K_L a$ vs h_r rotor submergence relationship it will be observed that for $Re < Re_{crit}$ the exponent of h_r is approximately 1.31, while for $Re > Re_{crit}$ it is round 1.0, implying the probable existence of a transition range, where the exponent 1.31 changes gradually to 1.0.

Under conditions where $Re < Re_{crit}$, the exponent obtained in the $K_L a$ vs rotor speed and in the $K_L a$ vs rotor diameter relations were 2.93 and 2.55,

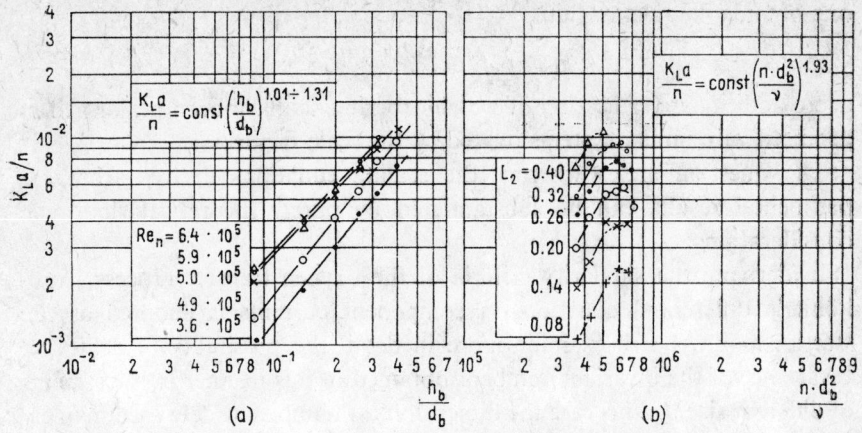

Figs 42(a) and *(b)*. Dimensionless plot of oxygen transfer variables

respectively. The effects of rotor speed and rotor diameter are normally taken into account by the peripheral velocity. This is why identical values are obtained for the two powers. However, investigations I have made have indicated the desirability of determining separately the relationships between the two variables and the $K_L a$ coefficient representing the rate of oxygen transfer. In addition to the preceding arguments this is justified by the following considerations. While it is recognized that the peripheral velocity is directly proportionate to both the diameter and the speed of the rotor, it is also found that their influence on the flow and oxygen transfer phenomena in aeration basins is different in magnitude. This will be seen by remembering that speed affects the inertial force transmitted to the flowing system, whereas an increase in the rotor diameter entails—assuming the same submergence depth—an increase in the inertial and frictional forces alike, since a rotor of enlarged diameter and submerged to the same depth presents a larger contact surface to the flowing medium.

According to Fig. 42a, in the range $Re > Re_{crit}$ the ratio $K_L a/n$ varies inversely as the Reynolds number. This phenomenon has not been observed in my experiments.

(c) Vertical-shaft units [182]

The power product of Eq. (269) is adopted as the similarity criterion of the oxygen transfer process and rewritten into a single dimensionless group. Introducing the scale factors and the conditions

$$\lambda_{h_r} = \lambda_{d_r} = \lambda_h = \lambda = 2 \qquad \lambda_g = \lambda_v = 1$$

one obtains the relationship

$$\lambda_{K_La} = \lambda_{d_r}^{0.41}\lambda_n^{1.33} = 1.33\lambda_n^{1.33} \tag{272}$$

as an additional verification of which experiments have been performed in geometrically similar systems related by the scale factor $\lambda = 2$.

The question arises whether the scaling equations derived from experimental results can be substantiated by purely theoretical similitude considerations.

In deriving the similarity criteria of the oxygen transfer process Damköhler's differential equation on component currents is adopted as the starting point. I will examine more in detail the possibility of achieving invariance of the invariant numbers, obtained in this manner in the systems of different size. In this case the dimensionless numbers widely used, namely the Ho, Pe, St and Da_1 numbers are included independently, without special power product combinations. Since no chemical reactions are involved the Da_1 number may be neglected. By applying similitude transformations the remaining three dimensionless numbers may be combined to yield the following similarity relationships [248]:

$$\frac{\lambda}{\lambda_t \lambda_v} = 1 \qquad \frac{\lambda_D}{\lambda \lambda_v} = 1 \qquad \frac{\lambda_{K_La}}{\lambda_v} = 1 \tag{273a–c}$$

or

$$\lambda = \lambda_t \lambda_v \qquad \lambda = \frac{\lambda_D}{\lambda_v} \qquad \lambda = \frac{\lambda_v}{\lambda_{K_La}} \tag{274a–c}$$

whence the set of equations

$$\lambda_t \lambda_v = \frac{\lambda_D}{\lambda_v} = \frac{\lambda_v}{\lambda_{K_La}} \tag{275a–c}$$

is obtained. From Eqs (275a–c) the following relationship may be derived

$$\lambda_t = \frac{1}{\lambda_{K_La}} \qquad \lambda_v = \sqrt{\left(\frac{\lambda_D}{\lambda_t}\right)} \qquad \lambda_v = \sqrt{(\lambda_D \lambda_{K_La})}. \tag{276a–c}$$

These equations in principle define the scaling relationships that can be derived from the Ho, Pe and St numbers.

In the above no attention has been given to the type or design of the structure and consequently the relationships derived are only applicable within the validity limits of Damköhler's equation. Now consider the implications of the scaling equations (276a–c), e.g. for the particular case of modelling oxygen transfer phenomena in INKA-type aeration basins.

The fundamental assumption is that in the case of aerobic fermentation and therefore also in activated sludge wastewater treatment, it is desirable for microbiological considerations to adhere to the condition $\lambda_{K_La} = 1$. In this way Eq. (276a) yields $\lambda_t = 1$, while Eqs (276b) and (276c) lead to the scale factor $\lambda_v = \lambda/\lambda_t = 1$ and in turn to $\lambda = 1$. It therefore follows that in the case of $\lambda_{K_La} = 1$ the effects of convection, conduction and mass transfer, i.e. the Pe and St numbers, can be taken into account simultaneously in a 1:1 model only. Just as in any other modelling problem the predominant dimensionless quantity must be identified, which may then be used as the basis of solving the scaling problem. In this particular case the choice is a fairly easy one, since the form of Damköhler's equation applying to aeration systems includes the dimensionless quantity

$$K_L at = \frac{St}{Ho}. \tag{277}$$

Returning now to the similarity criteria of Eqs (274a–c), it can be easily seen that Eqs (274a) and (274c), or the Ho and St numbers, play the dominant roles. Specifying again the criterion $\lambda_{K_La} = 1$, the scale factors $\lambda_t = 1$ and $\lambda_v = \lambda$ are obtained. These latter scale factors can also be used to calculate the scale factor of the air supply rate to be introduced into INKA-type aerators as

$$\lambda_{Q_{air}} = \lambda^2 \lambda_v = \lambda^3. \tag{278}$$

Allowance should, however, be made for the circumstance that the operation of INKA-type aeration systems is influenced by a number of variables (parameters of operation, boundary conditions, etc.), the effect of which cannot be entered directly into the computations in the above, relatively simple, scaling method. The approach suggested thus consists of using the empirical dimensionless relation describing the process studied for deriving the scale factor of the individual variables.

Equation (270) is in remarkably good agreement with Eq. (278) derived theoretically by equation analysis. At the same time Eq. (270) may be regarded an experimental verification of Eq. (278), since the empirical relationship has been obtained by processing experimental data.

3.4 INTERRELATIONS BETWEEN HYDRAULIC AND OXYGEN TRANSFER PHENOMENA

One should now ask the question whether it is possible at all to realize simultaneously the similarity of hydraulic and oxygen transfer phenomena. The investigations performed provide a positive answer to this question.

In Section 3.1 the validity of Eq. (263) has been substantiated experimentally with respect to INKA-type aerators, defining the scale factor of air supply rates to ensure similarity of hydraulic conditions. At the same time, similarity of oxygen transfer phenomena would require theoretically the use of the scale factor resulting from Eq. (278). The $\lambda_{Q_{air}}$ values obtained by the preceding two avenues for similitude should be equal to satisfy simultaneously the two criteria [248]

$$\lambda^2 \frac{w + v_k'' \lambda^{0.5}}{w + v_k''} = \lambda^3. \tag{279}$$

This equality, as will readily be seen, would require a 1:1 model. Equation (279) would imply further that the closer the scale factor to unity, i.e. the closer the agreement between the model and prototype dimensions, the smaller the difference between the similarity criteria of hydraulic and mass transfer phenomena will become.

As can be inferred from the above, simultaneous similarity of hydraulic and oxygen transfer processes is impossible to achieve, unless the model is a 1:1 replica of the prototype. For this reason model studies on these two processes are feasible by the method of partial scaling alone. Where the primary objective is to find the hydraulically efficient design of the structure the similarity criteria of flow conditions must be adopted, while for studying the oxygen transfer processes scaling must be performed, evidently, on the basis of the similarity criteria of mass transfer.

Repeated reference has been made to the specific relationships existing between the characteristic variables of hydraulic and oxygen transfer processes. The details of this problem have been examined at length in earlier papers [183]. Parallel to these experiments the similitude implications have also been investigated in order to determine the dimensionless groups which play major roles in the interrelations between these two processes. In earlier research work attention has been focused on the relationships between the operating parameters, such as h_r, d_r, n, etc., and the rate of oxygen transfer ($K_L a$, OC), these being of special interest to the designers [183, 248]. However, for obtaining a better understanding of the mechanism governing the mass transfer processes in aeration basins it was found necessary to investigate also the direct relationship between flow velocities and mass transfer rates.

As a result of similitude considerations made under the present research efforts the dimensionless groups derived from Damköhler's differential equation on component currents by similitude transformation analysis yield the following dimensionless expressions:

$$\frac{v^2}{DK_L a} = \frac{Pe}{St} = \frac{Pe^2}{Nu} = K. \tag{280}$$

The dimensionless number denoted by K, which represents a specific power product combination of the Peclet and Stanton numbers or the Peclet and Nusselt numbers, essentially comprises in dimensionless form the characteristic variables or parameters of convection (flow, v), conduction (diffusion, D) and mass transfer ($K_L a$) taking place in the system under consideration. Equation (280), when rearranged and letting $K=$ constant, yields [consistent with Eq. (276c)]

$$v^2 = K D K_L a \qquad (281a)$$

or

$$v^2 = \text{constant}_1 D K_L a. \qquad (281b)$$

For a particular flowing medium the diffusion constant D may be combined with the constant, so that

$$v^2 = \text{constant}_2 D K_L a. \qquad (281c)$$

This expression written for two different operating conditions (denoted by single and double primes) yields the fundamental relationship

$$\lambda_v^2 = \left(\frac{v'}{v''}\right)^2 = \frac{K_L a'}{K_L a''} = \lambda_{K_L a}. \qquad (282)$$

It can be inferred that—within a certain range—the velocities measured at discrete points within the flow field are related to the $K_L a$ values representing the rates of mass (oxygen) transfer in a way that the squared ratio of the velocities measured in the different systems equals the ratio of the corresponding $K_L a$ values.

The correctness of these theoretical considerations can be supported by data series obtained from experimental investigations on aerators of different type. The data available have been plotted in Fig. 43, by entering the v^2 data as ordinates with the $K_L a$ coefficients as abscissae. A definite relationship is seen to exist between the pairs of measurement data, which fitted with fair approximation to the linear relationship

$$v^2 = 67.5 K_L a - 165. \qquad (283)$$

Two types of deviation from Eqs (281a–c) can be distinguished:

(a) In the velocity range $v < 10$ cm/s the measurement points do not fit to the straight line plot. This is attributed mainly to the fact that in this range velocity measurements yield less accurate results.

(b) The fitted straight line does not pass through the origin, i.e. the abscissa $K_L a \neq 0$ pertains to the ordinate $v^2 = 0$. This deviation is assumed to be due partly to hydraulic conditions, partly to the circumstance that

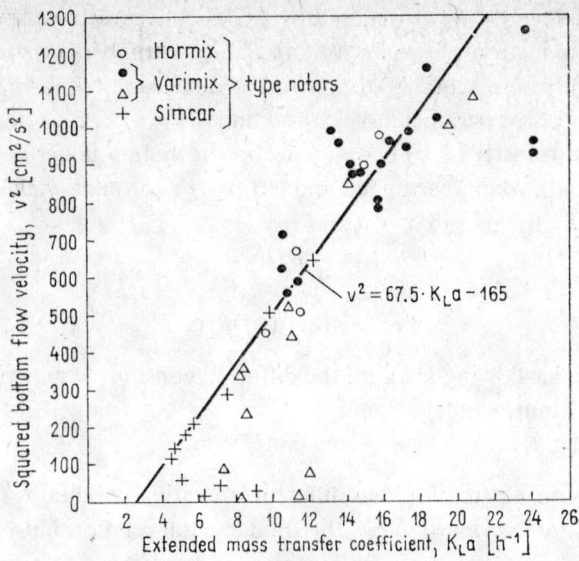

Fig.43. Flow conditions vs oxygen transfer in the aeration tank

some oxygen transfer occurs through the liquid–air interface even if the medium is stationary.

Attention is drawn here to the circumstances that could arise to give a misleading conclusion by neglecting that as a result of its dimensionless character Eq. (282) applies to different operating conditions in the same aeration basin or to corresponding operating conditions in geometrically similar basins. Neglect of these considerations could imply, for instance, that oxygen transfer conditions in a particular aeration reactor could be improved by a streamlined basin design in order to attain higher flow velocities. This, however, would be in conflict with practical experiences, as pointed out in earlier papers [183].

3.5 POWER CONSUMPTION

In a paper of mine published in 1966 I made an attempt at modelling approximately the power consumption of vertical-shaft aeration reactors [182]. I am perfectly aware of the formidable complexity of the problem and of the difficulties of establishing scaling relationships that could be applied to reproducing at least approximately the energy aspects of processes taking place in the systems of different size. Nevertheless, the approach suggested is believed to be correct as indicated by the desirability frequently experienced if comparing operating plant units of different size with each other. In such cases "model–prototype" relations may be examined in terms of similitude theory.

Parallel to deriving the empirical relationship given by Eq. (269) mentioned already, dimensional analysis and considerations of similitude have been applied in writing a dimensionless power product to express the power consumption of a novel type of vertical-shaft aerator reactor design [182]

$$\frac{N}{d_b^5 n^3 \varrho} = \text{constant} \left(\frac{h_b}{d_b}\right)^a \left(\frac{h}{d_b}\right)^b \left(\frac{n^2 d_b}{g}\right)^c \left(\frac{n d_b^2}{v}\right)^d \tag{284a}$$

or

$$Eu = \text{constant}\ L_2^a L_4^b Fr^c Re^d \tag{284b}$$

where N is the power consumption; and h, the depth of water in the aeration basin [the other notations have been explained in connection with Eq. (269)].

The measurement data processed by correlation analysis yielded the following empirical values: constant $= 6.499 \times 10^{-3}$, $a = 0.78$, $b = -0.27$, $c = -1.21$ and $d = 0.4$. During this experiment the circumstances were rather adverse for measurements to be taken, so that these empirical values should be viewed with reservation. Methodologically, however, I believe that the correctness of the similitude approach has been shown.

The line of reasoning outlined before has been stimulated by the scientific achievements on mixers reported by Rushton et al. [150], already referred to in connection with Eq. (170). As far as I know no similar measurements or data processing had been described specifically in connection with aeration systems at the time of this research effort (in the early 1960s). Subsequently Maise [140], Rácz et al. [181] and particularly Zlokarnik [180, 181] have made interesting contributions to this subject (see Section 2.3.2).

3.6 SIMILARITY OF REACTION KINETICS

Among the similarity criteria of reaction kinetics special importance is attributed in the case of activated-sludge systems to the various modified forms of the first Damköhler number [250]. My earlier investigations [251] have demonstrated the importance of the dimensionless sludge age S_a/t_c (where S_a is the sludge age and t_c the calculated average detention time), which can be shown to depend on two modified Da_1 numbers. The preceding time ratio yields with $\lambda_{t_c} = 1$, the scale factor $\lambda_{S_a} = 1$. Consequently, the sludge age in the systems of different size must be identical in order to approximate similarity of reaction kinetics. The importance of the dimensionless yield constant y is also emphasized (in analogy to the studies in the fermentation industry), which can be kept invariant in the systems of different size, resulting in a further refinement of accuracy in scaling reaction kinetics.

Reference is finally made to my earlier papers [251, 252] in which the similarity criteria of reaction kinetics in activated sludge systems have been evaluated in the light of actual data obtained from plant measurements.

3.7 BIOLOGICAL SIMILARITY

In an earlier paper [202] the question has been raised whether biological similarity can be defined at all. The conclusion arrived at was that the concept of biological similarity can be applied to the extent to which one succeeds in approximating or describing the biological processes by mathematical means. This must not be taken to imply that in the knowledge of the mathematical expressions describing a particular phenomenon (including also the initial and boundary conditions), similarity of different systems can be realized.

Available evidence supports the feasibility of describing by mathematical expressions some aspects of the biochemical and biological processes in wastewater treatment, and these expressions may be adopted in deriving the equations of similarity criteria.

Biological similarity can be achieved most conveniently by the method of trivial modelling [251]. As mentioned before, this involves realizing as identical as possible physical, chemical and biological conditions at the corresponding (homologous) points in the different systems, in which case the scale factors of the characteristic variables become unity. In this way the activated sludge substrate in a laboratory model may be considered substantially as a definite volume element of the prototype unit, the biocoenosis contained therein being "unaware" of the size, or even type, of unit housing it.

In ref. [202] I have also explained at considerable length the obstacles often encountered in complying with the criteria of trivial modelling. It should be realized that the physical (hydraulic), chemical and biological conditions which form a complex system are extremely difficult to reproduce. λ_x parameters in systems of different size raising the fundamental problems of trivial modelling. Irrespectively this method is commonly adopted in practice axiomatically, in general without offering theoretical justification. This approach may include several sources of error, especially when comparing data obtained in systems of different size.

Not even the efforts at achieving biological similarity by trivial modelling are believed to justify the neglect of similitude considerations in designing experiments and evaluating the data obtained. This assertion is supported by the foregoing considerations as well as a complex analysis of geometric, kinematic, dynamic and thermal similarity and by the similarity of mass transfer phenomena and reaction kinetics (which may be of primary interest in studying the unit operations in biological wastewater treatment). This may form the theoretical foundation—even in the case of trivial modelling—of comparing and evaluating the data obtained in systems of different size and scaling logical as well. Evidently, the scale factor λ_x of some variables will be unity, but others, such as operating parameters (rate of air supply, diameter and speed of the aeration rotor, etc.), will have scale factors other than unity even if the method of trivial modelling is used.

It will be pointed out subsequently that characteristic numbers, whether dimensional or dimensionless, can be determined by empirical techniques as well and may facilitate the realization or checking of biological similarity. Sludge age is quoted as a typical example as being closely related to the physical and particularly physiological properties of the activated sludge. The inclusion of sludge age in scaling considerations was also mentioned by Boyle and Rohlich [203]. The dimensionless sludge age defined by myself, which depends essentially on the modified first Damköhler numbers, can also be interpreted in this context.

An additional dimensionless number will finally be introduced which reflects a particular aspect of the biological process and may have implications in comparing biological systems. The familiar dimensionless relation

$$r_{i\max} t_g = \ln 2 \tag{285}$$

can be rewritten by the introduction of similarity transformation parameters into the form

$$\lambda_{r_i\max} = \lambda_{t_g} = 1. \tag{286}$$

Equation (286) is an expression of the fact that if Eq. (285) applies, the unit rates $r_{i\max}$ and generation times t_g measured in the single and double prime systems are in inverse proportion to each other.

The solution of scaling problems may be facilitated by the inclusion of the characteristic numbers mentioned above and others which may be conveniently defined.

3.8 THERMAL SIMILARITY

In view of the fact that the rates of chemical and biochemical reactions are controlled by temperature, the criteria of thermal similarity ($T'/T'' = \lambda_T$) should also be taken into account in scaling problems. In principle λ_T may assume values other than unity, but in the practical case of wastewater treatment systems adherence to the criterion $\lambda_T = 1$ is justified. In other words, the same temperature should prevail at the corresponding points in the different (single and double prime) systems.

3.9 TECHNOLOGICAL SIMILARITY

Consequent from the similitude principles outlined above, the similarity criteria related to the individual component phenomena of the technological process predominating in a particular problem must be taken into account. The set of similarity criteria characterizing the individual unit operations should then be examined in establishing the scaling and similitude equations of the treatment method. This line of reasoning prompted the introduction of the concept of technological similarity.

Beyond the scaling criteria considered so far it is possible to derive relationships that may be regarded as the typical condition equations of technological similarity. In this respect the known loading parameters may be adopted as the principal starting basis:

The scaling relationship of the volumetric or hydraulic loading is

$$\lambda_{T_h} = \frac{\lambda_Q}{\lambda_V} \tag{287}$$

the scale factor of the organic or BOD loading T_b is

$$\lambda_{T_b} = \lambda_{T_h} \lambda_{C_o} \tag{288}$$

the scale factor of the sludge loading T_s is

$$\lambda_{T_s} = \frac{\lambda_{T_b}}{\lambda_{G_L}} = \frac{\lambda_{T_h} \lambda_{C_o}}{\lambda_{G_L}} . \tag{289}$$

In the above expressions

Q is the wastewater flow; V, the net volume of the aeration reactor; C_0, the BOD_5 organic substance concentration in the arriving wastewater flow; and G_L, the sludge concentration in the aeration reactor.

In the interest of trivial modelling, identical loading conditions should be realized in the model and prototype

$$\lambda_{T_h} = \lambda_{T_b} = \lambda_{T_s} = 1$$

whence

$$\lambda_{C_0} = \lambda_{G_L} = 1$$

and

$$\lambda_Q = \lambda_V. \tag{290a}$$

Also, in geometrically similar systems

$$\lambda_Q = \lambda^3. \tag{290b}$$

The criterion $\lambda_{T_h} = 1$ will be noted to entail the same average retention times t and identical dilution rates $D = 1/t$, thus $\lambda_t = 1$ and $\lambda_D = 1$.

Operationally the activated-sludge aeration reactor and the final settling tank are organically interconnected by recirculation as an additional technological element, the scaling aspects of which will be considered in the following section.

3.10 THE ROLE OF RECIRCULATION

In the course of experimental investigations on wastewater treatment technology the question arises as to whether the recirculation parameters applied in the laboratory can be adopted without any further scaling as design criteria. The answer to this question is a negative one, supported by purely similitude considerations, but even more so in the light of experimental evidence.

As experimental evidence it will be recalled that, for instance, in technological tests at the laboratory recirculation rates of 200–300% or even higher are often found necessary in order to achieve satisfactory sludge concentration in the aeration tank. (The recirculation sludge concentrations G_R are usually lower in the model than in the prototype owing to the poorer sludge thickening effect of scaled-down settling tanks.) However, as indicated by actual plant experience recirculation rates of 25–100% ($R_Q = 0.25$–1.0) are normally acceptable, neglecting some special cases. For this reason it would be a fundamental mistake to adopt the R_Q value used in the model tests as design criteria uncritically. The economic implications of the problem, namely the costs of superfluous pumping, should be obvious.

The above arguments warrant further discussion of the scaling problem related to the conversion of recirculation variables from the model to the prototype. In the theoretical similitude analysis of the process, the invariance criteria of the pertinent equations will be adopted as the starting basis.

In earlier research work [253, 254] I succeeded in demonstrating the recirculation factor A introduced by Herbert [255] to be relevant in activated sludge systems as well

$$A = 1 + R_Q(1 - R_G) \tag{291}$$

where $R_Q = Q_R/Q$ is the discharge ratio of recirculation; $R_G = G_R/G_L$, the suspended content ratio of recirculation; Q, the raw sewage inflow to the plant; Q_R, the discharge recirculated; G_L, the concentration of activated sludge in the aeration basin; and G_R, the sludge concentration in the recirculated flow.

The factor A, as a dimensionless quantity, represents the rate of recirculation and can be demonstrated to control the equation of the invariance criterion

$$A = \text{idem} \quad R'_Q(1 - R'_G) = R''_Q(1 - R''_G) \tag{292}$$

implying the need for satisfying the condition $A' = A''$ in the different, corresponding systems, in the interest of scaling correctly the recirculation process. Equation (292) can be used to derive the fundamental relationship for the single and double prime systems

$$\lambda_{R_Q} = \frac{R'_Q}{R''_Q} = \frac{R''_G - 1}{R'_G - 1} = \frac{(G''_R - G''_L)G'_L}{(G'_R - G'_L)G''_L}. \tag{293}$$

Experimental results can now be entered into Eq. (293) to find the scale factor of R_Q. Also inherent in Eq. (293) is the realization that the equality $R'_G = R''_G$ is impossible to satisfy unless $R'_Q = R''_Q$. Consequently, the recirculation discharge ratio adopted in the model can only be transferred to the prototype if the sludge concentrations are identical. The sludge concentration in the recirculated flow is, however, normally much lower in the model than in the actual plant size so that the recirculation discharge rate is necessarily higher in the former.

No limitations having been imposed on geometrical similarity when writing Eq. (293), the relationship retains its validity also in the case of different, geometrically dissimilar systems. Consequently when comparing similar technological processes in two systems the quantities characterizing recirculation should be related as given by Eq. (293). The single and double prime systems may be a prototype and a model or the two may be of the same order but of different design or may operate with different plant parameters.

Scaling according to Eq. (292) will obviously result in similar conditions with respect to recirculation alone. Other similitude aspects of the process involve the observation of additional similarity requirements.

3.11 PARTIAL AND FULL SIMILARITY

It will already be appreciated that perfect similarity scaling of activated sludge systems is impossible to obtain. The validity of this conclusion can be demonstrated analytically for the general case. For this reason partial similarity and, in turn, only approximate modelling can be attempted [248, 251]. Nevertheless, experience has shown the method of partial modelling to yield acceptable results in the majority of problems encountered in practice.

The previous considerations imply further that in particular cases it may be convenient to resolve the overall modelling problem into several part problems. The aerator-settling tank system considered before may thus be reproduced, for instance, in the following steps:

1. Modelling the aeration reactor by similarity of the
   (a) hydraulic,
   (b) mass-transfer,
   (c) reaction kinetic, and
   (d) biological processes.
2. Modelling the settling tank by similarity of the
   (a) hydraulic, and
   (b) settling processes.

This subdivision is an arbitrary one, but it includes the main unit operations of the treatment technology. In any particular case an analysis of the problem will reveal which aspect of the complete process is to be regarded predominant and also the similarity requirements to be observed.

3.12 INITIAL AND BOUNDARY CONDITIONS

As stated by the third principal similitude theorem, the theorem of Kirpitshov–Guhman, in establishing the necessary and adequate criteria of similitude the initial and boundary conditions must also be allowed for besides the equations describing the processes and the invariance requirements. These requirements for non-ambiguity can be determined in the knowledge of a particular problem only. Attention will therefore be directed to a few considerations and examples of more general interest [251].

The wall effect is more or less pronounced in virtually all cases and may modify not only the hydraulic but also the physico-chemical and biological conditions. The boundary surfaces may, for instance, act as catalyser and the biological film developing on the wall must also be taken into account —especially in small tanks. The ratio of the internal (active) boundary surfaces and volumes of the reactors of different size should be possibly equal. In geometrically similar units, where the surfaces and the volumes are proportionate to λ^2 and λ^3, respectively, the preceding surface : volume ratio is proportionate to λ^{-1}. This implies that the wall effect is more important in smaller units. For this reason it may be advisable to inactivate the wall of laboratory models.

As an additional example the role of the settling tank is mentioned which must be taken into account as an external or boundary condition in studies on aeration reactors. A requirement in modelling such systems is that the activated sludge recirculated from the secondary settling tank to the aeration basin should have identical properties in the model and prototype. Laboratory models of settling tanks are therefore liable to be over- or underdimensioned if the volume adopted is too large or too small. Experiments with unmatched units (e.g. aeration basin and settling tank) cannot be justified even if a saving in cost is attained by the use of existing equipment. The model settling tank should operate to satisfy the scaling requirements of the aeration reactor.

Additional examples can be mentioned for taking the initial and boundary conditions into consideration when specifying the criteria of similitude and scaling. The variations in the sewage flow to the structures, further in the quality of sewage, etc., are among the factors to be allowed for in modelling. Experience gained in experimental work has demonstrated the usefulness of observing the following requirements when scaling activated sludge systems.

Care should be exercised first of all to adjust the initial and external (boundary) effects influencing the systems considered in a manner consistent with the above requirements. Identical times, temperature and especially concentration conditions are essential in the model and prototype. Should a dominant quantity, e.g., the concentration parameter C, prove for one reason or another impossible to obtain with the same value, i.e., at $\lambda_C = 1$, then the effect should be compensated for by some means. This is well exemplified by the case of recirculation, where—as already mentioned—different sludge concentrations are in general inevitable in the model and the prototype.

3.13 SCALE EFFECT

In various scaling problems a limit value is observable in one of the significant dimensions where the process studied undergoes a substantial change. This effect, commonly called the scale effect, may appear in the broader sense not only with a single significant dimension, but also at a certain limit value of any variable.

In a paper referred to before [202], I have mentioned several examples of the scale effect observed in the unit operations of activated sludge treatment. The wall effect discussed in this paper may also be classified among these. In this respect an additional example will be introduced here.

The method of trivial modelling involves the requirement that the activated sludge flocs present in the treatment system should be of the same size (in the statistical sense) in the units of different size. Attention has already been given to the importance and implications of this requirement elsewhere [245] since the scale effect related to floc sizes may be an appreciable one. Assume that the floc size is reduced in compliance with the geometric similarity requirement. This entails a substantial distortion, since the molecular sizes and even the size of the bacteria are fixed. At the same time the change affects the specific surface of the sludge flocs which plays a significant role in mass transfer *via* diffusion. Consequently the oxygen supply to the flocs (the aerobic, or even anaerobic conditions within the flocs), and the removal rate of the products formed in the course of assimilation suffer fundamental changes as a consequence of the change in scale, representing a typical example of the scale effect. This phenomenon has been pointed out by several authors—in a more pronounced form by Pasveer (see Boyle [203,203a]) — though not on the basis of similitude considerations. Hinshelwood [256] emphasized by physical (heat transfer) analogy, specifically with reference to bacterial cells, the role of the scale effect which may influence fundamentally even a physiological phenomenon like cell division by modifying the cellular concentrations.

In analysing the advantages and drawbacks as well as possibilities and limitations of laboratory, pilot and plant-scale investigations, Märki (see Horváth [291,256a]) adopted a practical approach to the scale-effect problem. Besides the numerous advantages (e.g., from technological aspects) obtainable by using laboratory-size equipment, he emphasized the need of remembering the difficulties of reproducing the actual effects of plant operation, such as the role of quantity and quality fluctuations. In agreement with myself he expressed the inadvisability of extending the cost data of laboratory model measurements to scaled-up plant equipment, especially if the models are very small in size. For such purposes the data should be obtained from at least pilot-scale equipment.

3.14 COMMENTS AND DISCUSSION

The comments, contributions and reviews made regarding papers I have published over the past 15–20 years on the applications of similitude in wastewater treatment will subsequently be discussed.

The comments on the paper [173] submitted to the 1966 International Conference of IAWPR (Munich, F.R.G.) will be dealt with first. In his comments Professor Dobbins (New York University) emphasized the desirability of developing dependable scaling methods in the field of wastewater treatment [257]. In his opinion the effect of the Fr and We numbers should also be taken into account when writing the dimensionless relations defining the extended mass transfer coefficient $K_L a$. On the paper [173] he commented: "The experimental data provide a good fit to the functions proposed by the author and the ranges of validity have been stated clearly." In connection with the paper of Kalinske [162] he too mentioned the scaling problems related to the coefficient α_{OC}. I completely share his views.

Concluding their comment Rousse and Brouzes (Paris) stated [258]: "A methodical organization of the tests and a rational exploitation of the results in the way shown by Horváth should allow an increase in the exactness and amount of information drawn from the studies relating to the transfer of oxygen."

Inoue (Kyoto University) compared the results of oxygen transfer experiments performed with porous nozzles in a deep injection basin with my data obtained in an INKA-type basin [259]. The significant difference in the geometrical design of the two aeration basin types was shown to cause— as expected on the basis of theoretical considerations—appreciable differences in the expressions describing the oxygen transfer conditions. These investigations demonstrated therefore the importance of geometric similarity.

In his verbal comments Möller (Munich) analysed the problems associated with the scaling of the oxygen saturation concentration [260]. Starting from this he questioned the possibility of modelling oxygen transfer processes. These reservations are believed to be justified only if one wants to realize perfect similarity. However, as demonstrated positively by a number of scientific results (cf. Section 2.3.2) as well as by my investigations, partial similarity meeting technological requirements can be obtained in practice.

Wilderer and Hartmann (Karlsruhe) agreed with the conclusions of the paper [173] that the inclusion of the Re number is essential in the dimensionless relations describing the oxygen transfer and uptake processes [261]. They compared the relations I suggested, which correspond formally to the correlation of Gilliland–Sherwood, with Frössling's dimensionless relation and found the latter to have a wider range of validity. Incidentally, to my knowledge Frössling's equation was first used by Hammerton and Garner [262] to describe oxygen transfer processes in fine-bubble aeration systems.

Wilderer and Hartmann considered the special case where $Re=0$, for which the Gilliland–Sherwood relation would yield $K_L=0$, which of course does not actually hold, since mass transfer occurs in stationary media as well, again demonstrating the paramount importance of defining exactly the validity range of the power product dimensionless relations. Parallel to these considerations they emphasized the important advantages inherent to the dimensionless relations written in the form of power products and with clearly defined ranges of validity, in that the invariance is more readily ensured in similarity transformations. This in turn is of paramount importance in modelling, specifically in determining the scaling criteria. The foregoing conclusions are consistent with the principles explained in connection with Eqs (131) and (132) in Section 2.2.3.

The journal *Das Gas- und Wasserfach* [263] presented a comprehensive review of the paper on hydraulic model testing, published in *Hidrológiai Közlöny* [243]. In the course of the discussion on Eckenfelder's paper Kalinske [264] quoted some of my conclusions [182] which were in agreement with his. The data related to the oxygen transfer and power demand of vertical-shaft aeration systems.

More recently Schmidtke (University of Waterloo) described at considerable length [171] my investigations on the modelling of aeration units emphasizing the importance of similitude theory in scaling problems. Employing the methods for modelling wastewater treatment equipment he published interesting experimental material on this subject, as referred to in Section 2.3.2.

A joint paper [172] submitted to the 1976 Conference of IAWPR (Sydney) discussed under a novel approach—founded on detailed experimental work — the scaling methods of vertical-shaft aeration reactors.

In his investigations on vertical-shaft aeration systems Rácz (Netherlands) adopted—with reference among others to my paper, methods of research and data processing [265] resembling those described in the comprehensive monograph published by VITUKI [183]. I, at the same time, presented a clear review of the application of dimensional analysis in this particular domain (stirring, mass transfer).

Following an invitation by the organization committee, I presented a paper at the 1978 IAWPR Conference (Amsterdam) on scaling-up problems associated with aeration systems [266, 267]. In his comments on the paper Clark (United Kingdom) concurred with me in emphasizing the conclusion that any application of similarity theory should be founded on practice [268].

Reference is made again to the paper of Zlokarnik published in 1979 [180], in which he reviewed comparisons of data from the literature, also mentioning the paper of Schmidtke and Horváth [172].

In Hungary, Salamin presented a brief review [269] of the model studies on air injection systems. In his comments on the paper [270], Benedek [271] suggested the establishment of a working group to develop the methodological problems related to the scaling of wastewater treatment units. Hock [272] called attention to the importance of processing statistically the data obtained by scaling methods.

4. Economic similarity

4.1 FORMULATION OF THE PROBLEM

In designing wastewater treatment plants and their various component structures problems are frequently encountered which can be related to scaling up or down the dimensions or capacities of the structures. As a specific example reference is made to the rather coarse estimates in which the capacity and cost data of a project constructed earlier are used in relation to a contemplated project of smaller or higher capacity. This virtually implies cost estimates or predictions based on unit costs.

Having examined the costs and cost functions of a variety of structures and plants used in wastewater treatment—and reported in several publications—the following conclusions have been arrived at:

(a) The dimensions and capacity of structures and plants are closely related to the costs of construction. If a cost function of particular form is assumed to apply, the values of the constants involved therein may vary—of course beyond certain capacity limits. Consequently, the shape of the specific cost function applying to a particular case may be influenced by increasing or reducing the capacity.

(b) From a number of cases reported in the literature it is found that although the costs increase in absolute magnitude as the capacity is increased, the unit values related to capacity show a decreasing trend (degressive cost). The formulation normally adopted in practice is that the construction of plants of higher capacity is in general more economical, provided of course that the increment capacity is actually used.

(c) The considerations outlined above may be generalized to a certain extent on the basis of scaling considerations. This generalization leads, in turn, to the similitude interpretation of the problem.

The aim of this chapter is to analyse the problems of economic similarity and capacity scaling associated with the construction of structures and plants in wastewater treatment, starting from cost functions of specific shape.

In this manner an estimation method is suggested, in which allowance is made for the cost factors. The method consists essentially of extrapolation based on similitude transformation.

4.2 REVIEW OF LITERATURE

Scaling problems of wastewater treatment facilities with due regard to economic aspects have already been studied by some authors. This problem arises mainly in connection with the optimal scheduling of projects, namely determination of the optimal alternative of projects constructed in several stages. Investigations in this domain were founded in process engineering on scaling studies related to chemical plants. The economic aspects of scaling were studied in the fundamental paper of Dickerson [273] who analysed the scale-up possibilities of individual unit operations as well as complete plants in the chemical industry. Reference is made further to the paper by Walley and Robinson [274] who established relationships between plant sizes and the economics of production in the petrochemical industry by applying optimization considerations. The main conclusions of these papers are also of interest in the field under consideration.

In the literature on the subject specific reference to wastewater treatment projects is found in the papers of Tchobanoglous [275] and Rachford *et al.* [276]. The latter developed an optimization method based on cost functions for different types of water and wastewater treatment projects. Of the Hungarian literature mention is made further of the paper by Ábrahám and Takács (see Horváth [291,276a]) in which a graphical method is presented for the optimal scheduling of communal treatment plants in connection with a partially completed wastewater treatment project. Using a scaling method based on similitude considerations, I have also suggested a scale-up procedure in a paper published in 1978 [266]. The ideas presented there will be expanded in more detail subsequently. Reference is made at the same time to Section 2.4.1 where a practical example on digester scaling has also been described.

4.3 SIMILITUDE AND SCALING CONSIDERATIONS

Model studies based on similitude theory are frequently applied to hydraulic and other phenomena, while in the analysis of economic problems mathematical models are more common. These two avenues of thought are believed to supplement each other conveniently within certain limits as can be demonstrated in this particular field.

The problem normally encountered in model testing consists of transferring (converting) the data obtained from laboratory or pilot experiments to prototype units. In the majority of cases this involves scale-up, but in special cases the problem may be one of scaling down. In another case extrapolation based on similitude transformation is involved, offering the possibility of establishing mutual and positive relationships between the processes and phenomena taking place in the systems of different size.

Viewing the concept of modelling from a more general aspect a particular prototype unit may also be regarded as the model of another full-size unit. This kind of interpretation is possible mostly in the case of systems differing in size. Comparisons between systems of identical size are also possible, for instance if the technological goal specified (e.g. a given treatment efficiency) is realized by methods or means differing from the first. These latter are the alternatives which may serve as the basis of economic comparison between different systems.

In order to avoid misinterpretations it should be emphasized that, in general, serious difficulties are encountered in converting parameters of economic relevance from the laboratory model to prototype size. No solution of this kind can normally be expected to produce acceptable results, in contrast to the scaling solutions in which the model–prototype conversion yields fairly approximating and acceptable results, for instance from the viewpoint of hydraulics.

While emphasizing the foregoing considerations, extrapolation in the domain of economic analysis is believed to be fully justified, provided that certain conditions are satisfied and the methods used are founded on similitude theory. Starting from given cost functions a convenient approach will be outlined subsequently for the solution of this problem.

A cost function valid in a particular range shall be assumed to have the form

$$K = \text{constant } C^a \qquad (294\text{a})$$

where K is a characteristic cost (cost function or operation); and C, a variable representing the capacity of the structure or plant.

Expressing capacity by the arriving wastewater flow, this becomes

$$K = \text{constant } Q^a \qquad (294\text{b})$$

where Q is the mean daily flow, for example.

Consider hereafter two alternatives (systems) which are to be related to each other. (The quantities in the two alternatives will be distinguished by single and double primes.) From Eq. (294b) one may write for the two alternatives the following relationships

$$\frac{K'}{K''} = \left(\frac{Q'}{Q''}\right)^a \quad \text{or} \quad K' = K'' \left(\frac{Q'}{Q''}\right)^a. \tag{295a,b}$$

Introducing for the ratios of the corresponding quantities the similarity transformation parameters (the scale factors) one obtains

$$\lambda_K = \lambda_Q^a. \tag{295c}$$

The scale factor λ_Q may also be written in terms of the corresponding volumes V and times t as

$$\lambda_Q = \frac{\lambda_V}{\lambda_t}. \tag{296a}$$

For instance, if $\lambda_t = 1$ (related e.g., to one day), then

$$\lambda_Q = \lambda_V. \tag{296b}$$

From the preceding expressions

$$\lambda_K = \lambda_V^a \quad \text{or} \quad \frac{K'}{K''} = \left(\frac{V'}{V''}\right)^a. \tag{297a,b}$$

Consider the variations in unit costs again understood as the costs related to unit capacity

$$\frac{K}{C} = \text{constant} \quad \frac{C^a}{C} = \text{constant } C^{a-1}. \tag{298a}$$

Substituting the flow rate Q or the volume V for capacity, this becomes

$$\frac{K}{Q} = \text{constant } Q^{a-1}; \quad \frac{K}{V} = \text{constant } V^{a-1}. \tag{298b,c}$$

The criterion $a = 1$ may be regarded a special case, where

$$\frac{K'}{Q'} = \frac{K''}{Q''} = \text{constant} \tag{298d}$$

or

$$\frac{K'}{V'} = \frac{K''}{V''} = \text{constant}. \tag{298e}$$

These expressions convey several implications of practical significance concerning the extrapolation of cost data to systems of different size. Evidently, the prerequisite is the validity of a cost function of the type described by Eqs (194a, b). However, before embarking upon a more detailed discussion of these implications it is necessary to define in technico-economic terms the experimental constants involved in these relationships.

4.4 INTERPRETATION OF THE CONSTANT AND THE EXPONENT a

It should be emphasized first of all that the constant values involved in the above relationships (including also the cost function adopted as the starting basis) may be regarded constant within certain ranges only. On the other hand, the constant quantities in the individual expressions are not necessarily identical [e.g. the constant in Eq. (294a) must not be identified with that in Eq. (298a).]

A few remarks should be added here concerning the interpretation of the constant and the exponent a. In Eq. (294a) adopted at the outset $K=$ constant if $C=1$, implying that this constant represents the cost pertaining to unit capacity. Formulated in a different way this means that the value of the constant includes the effect of all main factors influencing the cost K, which actually control the costs besides the capacity C. Nothing else is implied but that under the given conditions all relevant quantities other than the independent variable C are assumed constant in the particular case. In this formulation the constant may be understood as a constant multiplier of the cost function.

In the present approach the exponent a plays a role of fundamental importance. As will be shown subsequently the magnitude of a may also be related to scaling. In this respect reference is made to Eqs (295a–c) and (197a, b). With regard to the foregoing, the exponent a has a constant value applying to a particular case. It must be determined empirically and may be termed the scale-up factor.

Concerning its numerical value, from the analysis of cost functions applied in practice and examined here, the exponent a may be classified into two main ranges:

(a) In the majority of cases of practical interest a assumes positive values smaller than unity (decreasing costs), the most frequent values being situated between 0.5 and 0.9.

(b) In special cases a may assume values larger than unity (increasing cost). Reference is made here to the papers of Negaard [277] and Tihansky [278].

(c) The case $a=1$ may be mentioned as the boundary between the preceding two ranges and is also of definite practical significance (proportionate costs). As a theoretical example Eqs (298a, b) are mentioned, while a practical case is described by Negaard in the paper previously referred to.

4.5 PRACTICAL DETERMINATION OF THE CONSTANTS INVOLVED IN THE COST FUNCTIONS

Although the similitude aspects of the problem under consideration have been left unmentioned in the professional literature, the efforts aimed at the determination of the constants involved in the cost functions provide excellent material for the application on the approach suggested, namely the method of scaling or extrapolation. The actual data published in the literature can be conveniently used for this purpose. Some of the data will be presented later in tabular form.

Values of the exponent a have been compiled in Table 8 from the papers

Table 8

Values of the exponent a in different treatment facilities
(after Rachford *et al.* [276])

Type of treatment unit	Exponent a
(a) Small treatment plants (0.05 to 1.0 mgd)	
Imhoff tank	0.54
Rotating disc aerator	0.62
Trickling filter	0.62
Activated sludge system	
with separate digester	0.62
with integrated sludge treatment	0.67
Stabilization ponds (lagoons)	0.60
Land disposal (wastewater irrigation)	0.58
(b) Medium and large plants	
Imhoff type plant	0.757
First treatment stage with separate sludge digester	0.641
Stabilization ponds	0.654
Activated sludge plant	0.798
Trickling filter with separate sludge digester	0.648
Trickling filter combined with Imhoff tank	0.727

published on different treatment plants and structures. Reference is made in this context also to the papers by Negaard [277], Heberling and Hahn [279].

As already mentioned the exponent a may, in special cases, assume values larger than unity, though normally the reverse is true. A few examples of this case will also be given.

At plants treating industrial wastewaters Tihansky [278] found *a* values higher than unity. In this particular case he related the unit value of annual production in U.S. $/ton units to the daily effluent discharge in the form of a cost function. Cost data on sludge treatment were processed by Negaard [277] who found also an exponent higher than unity. In this particular case the costs of sludge treatment and disposal were related to the *per capita* equivalent expressed as hydraulic loading. The calculations yielded $a=1.25$.

4.6 POSSIBILITIES OF FURTHER GENERALIZATION

From the professional literature on the analysis of cost functions related to wastewater treatment facilities it is inferred that some authors have made successful attempts at going beyond the capacity variable additional parameters in the cost functions. The use of cost functions written in power product form appears to be especially important, as will be recalled; in such cases the employment of similitude transformation is readily solved in scaling and extrapolation.

The use of power product cost functions was suggested among others by Negaard, emphasizing that in this way the constants involved in the cost functions are fairly easy to evaluate, for instance by graphic curve fitting in a log–log system of coordinates. Mention is also made of the comprehensive paper by Tihansky who included in the relationship derived, in addition to the *per capita* equivalent representing capacity, the BOD concentration of the raw wastewater introduced to the treatment plant. A similar approach was adopted also by Young and Carlson [280]. The latter derived a power product expression and determined the constants involved experimentally

$$K = \text{constant } Q^a C_e^b. \tag{299}$$

In the foregoing expression

K is the total annual cost; Q, the average daily inflow (mgd); C_e, the concentration of pollutants in the treated effluent (mg/l); and a, b, exponents determined by the evaluation of experimental data.

In the case of conventional activated sludge biological wastewater treatment the specific values determined by Young and Carlson were

$$a = 0.7072 \text{ and } b = -0.3256.$$

Evidently, Eq. (299) can also be used within the corresponding validity limits to determine the relationship of the scale factors (similitude transformation parameters). Assuming that the particular constant has identical values in the single and double prime systems, one may write

$$\lambda_K = \lambda_Q^a \lambda_{C_e}^b. \tag{300a}$$

As special cases the following alternatives may be considered for $\lambda_{C_e} = 1$, and Eq. (300a) is simplified to Eq. (295c). Moreover, if $\lambda_K = 1$, then

$$\lambda_Q = \lambda_{C_e}^{-b/a}. \tag{300b}$$

This expression may be used in the case where it is specified that the costs be identical in the single and double prime systems, but the scale factors of the inflow to the plant and of the pollutant concentration in the treated effluent have values other than unity.

4.7 ECONOMIC SIMILARITY

The first question to be answered is whether the extension of similitude methods to the domain of economic analysis is justified or not. In my opinion both the theoretical and the practical foundations may be regarded as realistic concerning this methodological approach. The results and techniques of model tests based on similitude theory as well as cost calculations founded on mathematical models may supplement each other effectively. Moreover, the fact that the methodological appraisal of various technological processes and operations and the economic calculations may be founded—to a certain extent and within certain limits—on common theory and methodology may also be regarded to favour a similar approach.

Parallel to these considerations it should be emphasized, however, that the concept of similitude is interpreted to a certain extent differently from the formulation employed in connection with technological processes, such as the model study of hydraulic operations. The exact requirements and implications of geometric similarity in the evaluation of certain model tests are mentioned in this context. Of course, it is also realized that the requirements of geometric similarity can be violated in certain cases.

In the concept of economic similarity the ratios of the corresponding cost data and of the characteristic quantities controlling these play a decisive role in determining the relationship between the similarity transformation parameters and in extrapolation calculations.

When comparing the economics of systems of different size, for instance plants of different capacities or units of different dimensions, the relations of quantities representing particulars of technological processes or operations are usually neglected. Essentially the same line of reasoning is followed, e.g. when the criteria of hydraulic similitude are observed while the criteria of thermal similarity or similarity of molecular processes are neglected. The basic philosophy is in each case to start from the relationships determining or describing the basic features of the problem and to determine therefrom—observing the relevant initial and boundary conditions—the relation or set of relations describing the mutual and unique interrelations between the systems of different size (e.g. as the relationship of the relevant similarity transformation parameters). This approach is fully consistent in its substance with the concept of the invariant function introduced earlier. In the case under consideration the cost function may be regarded as the invariant function and may actually be found invariant during similarity transformation. Evidently in particular cases it may be found necessary to substantiate the invariant nature of the function.

As already mentioned, the concept outlined is not primarily concerned with the extrapolation of data determined in laboratory-scale models. Practical experience in wastewater treatment has revealed that the economic parameters determined in excessively scaled-down systems cannot be transferred with an accuracy acceptable for practical purposes to prototype-scale equipment. For this reason the approach in question is considered primarily applicable to the extrapolation of data determined in pilot or full-scale plants. Within these limits it is logically possible to investigate scaling up or down and increasing or reducing capacity. This latter distinction is considered appropriate since in certain cases (e.g. structures) capacity can also be expressed in terms of the net volume of the structure. In other cases, however, it may be found desirable to express capacity in terms of other characteristic quantities. For such purposes, e.g. the number of population served, the *per capita* equivalent, the raw wastewater (sewage) discharge or other index numbers may be used.

It will be appreciated that the concept of the scale effect may also enter the field of economic analysis. It should be noted in this context that extrapolation beyond certain scale limits cannot be expected to yield acceptable results. This is due mainly to the fact that the cost functions retain their validity within a certain range only. It should be realized moreover that by generalizing the form and validity of a particular cost function the constants involved may also assume different values.

Starting from similar considerations one may be justified in using instead of the conventional "scale effect" term the variability of the relationships related to the system examined. It should be noted that conclusions of this nature have already been derived in my papers reporting on earlier research work concerning the generalization of similitude methods. The characteristic relationships may thus change in form without appreciable differences in the dimensions of two or several facilities compared. Unit operations realized at treatment plants of almost identical dimensions with minor differences in the design of structures, in technologies and treatment efficiencies may be mentioned as examples in this context.

4.8 CONCLUSIONS AND RECOMMENDATIONS

(a) In analysing the economics of wastewater treatment plants and structures it is considered desirable to introduce the scaling methods applied successfully in several fields of engineering science, founded to a certain extent on similitude theory. In this way extrapolation can be made more dependable as regards both scaling up and down of dimensions and capacities.

(b) Extrapolation is founded on the cost functions established in the course of the economic analysis. In order to make similitude transformation actually applicable, the cost functions are written most conveniently in the form of power products. The practicability of this approach, i.e. the applicability of product expression as cost functions related to wastewater treatment facilities, is corroborated by the results of analyses described in the literature on the subject. It should be emphasized, however, that the reliability of any scaling conclusion is confined to the validity range of the fundamental relationships.

(c) In the extrapolation of data the exponent a involved in Eqs (294a, b) assumes a special significance and may be termed the scale-up factor. The magnitude of a ranges normally for different wastewater treatment facilities from 0.6 to 0.8 (3/5–4/5)—as evidenced by an extensive survey of the literature. As a first approximation the value 0.7 ($\sim 5/7$) may be assumed for rough information. For more accurate calculations a can be evaluated from the data available.

(d) Equations (295), (298) and (300) can be used for performing extrapolation in particular cases.

(e) The analytical approach, in which the unit costs are used to estimate directly the costs of equipment, amounts essentially to the assumption of $a=1$, in which case the relations of Eqs (298a, b) are applicable.

(f) In view of the fact that in practice the value of a is usually smaller than unity, any cost estimate based on the unit costs mentioned before is liable to be misleading. Consider, for example, the case where the unit cost of a structure is known and expressed in $/m³ net volume units. The costs of a similar structure would be obtained by simple multiplication with this unit cost. This method of estimation is known to be rather widely used. The similitude considerations have, however, revealed some of the serious shortcomings and limitations inherent to this approach.

(g) It should be added finally that the arguments outlined before suggest the application of methods and techniques related to systems engineering and operations research.

5. General conclusions

The conclusions arrived at and the results obtained in connection with the specific processes have been summarized in the respective chapters. Some of the more general principles and conclusions only will be recapitulated here [281–284].

As demonstrated by the wealth of evidence presented in the literature on the subject, as well as my experimental results, the technological processes in wastewater treatment systems can be scaled up with a fair degree of approximation by taking the main operational variables and parameters into account. There is also ample evidence available—consistent with earlier theoretical considerations—to indicate that unless supported by similarity and scale-up analyses any model study into wastewater treatment operations is susceptible to misinterpretation and that the data of laboratory investigations are liable to result in misleading design criteria.

Modelling without scale-up considerations, as widely adopted in wastewater treatment practice, is believed to provide no more than qualitative answers to particular problems, for example whether a particular effluent is degradable or not, while the quantitative, specific information is much less reliable. An approach based on similarity theory can certainly be expected to yield much more accurate design information and criteria.

The relationships and conditions found essential in earlier investigations for proper modelling have been presented in this book. For example, in scaling up activated sludge systems it is considered essential to ensure identical times, loads per unit volume, concentrations, temperatures and reactivity conditions in the different scales in order to reproduce by the model and the process taking place a definite volume element of the prototype unit.

These investigations confirmed that perfect similitude of model and prototype is impossible to achieve. Nevertheless, by introducing in particular cases the predominant characteristics and conditions, by adopting the technique of "partial modelling", it is possible in general to reproduce the technological processes and to attain in this way "technological similarity" of an accuracy adequate for practical purposes.

Experimental devices of different size (laboratory, pilot and prototype scale) are normally used in research on problems related to the technology of wastewater treatment. It should be realized, however, that the processes take in general different courses in the units of different size so that the corresponding variables may assume different numerical values. This is the reason justifying the efforts at establishing mutual and positive relationships between the processes which take place in the model and prototype systems, with the aim of deriving reliable information from one system on the other (by scaling up or down).

The number of publications on the problems of similitude and scaling-up is rather small in the literature on wastewater treatment. Much of the information is presented rather in the form of brief references included in detailed descriptions of particular technological processes or equipment. In view of the both theoretical and practical importance of the question much greater attention is believed to be warranted to the solution of similitude and scale-up problems in the field of wastewater treatment. The results achieved in hydraulics and chemical process engineering specifically in the fermentation industry [284–287] may offer welcome guidance in this respect.

Future research should be oriented at extending the sphere of proper scale-up criteria, at improving the accuracy and especially at the experimental determination of their ranges of validity.

Literature

1. Kirschmer, O.: *Mitteilungen des hydraulischen Institutes der TH München* **I**, 21 (1926).
2. Horváth, I.: Mechanikai szennyvíztisztítás. (The primary stage of wastewater treatment) VMGT 50, VIZDOK, Budapest 1973.
3. Geiger, H.: *Sandfänge für Abwasserkläranlagen.* Archiv für Wasserwirtschaft. Maschinenfabrik H. Geiger, Karlsruhe 1942.
4. Horváth, I. and Tasfi, L.: *Vízügyi Közlemények* **3**, 468 (1964).
5. Horváth, I.: Válogatott fejezetek a szennyvíztisztítás és a vízelőkészítés köréből. BME Továbbképző Intézet kiadványa (Selected chapters in the field of water and wastewater treatment. Publication of the Institute of Extension Courses of the Technical University of Budapest) Budapest 1970.
6. Horváth, I.: Légbefúvásos homokfogó vizsgálata. VITUKI kiadvány (Investigation of aerated sand trap. Publication of VITUKI) No. **38**, Budapest 1972.
7. Sallay, K.: *Hidr. Közl.* **7**, 324 (1968).
8. Albrecht, A. E.: *Water and Sewage Works* **9**, 331 (1967).
9. Rubey, W. W.: *Am. J. Sci.* **25**, 19 (1933).
10. (Lyashchenko, P. V.) Лященко, П. В.: Гравитационные методы фазовой селекции. Гостоптехиздат, Москва 1940.
11. Field, W. G.: *Journal of the Hydr. Div., Proc. ASCE* HY3, 705 (1968).
12. Bewtra, J. K.: *Water and Sewage Works* **2**, 60 (1967).
13. (Shifrin, S. M.) Шифрин, С. М.: Современные методы механической очистки сточных вод. Стройиздат, Москва 1956.
14. (Arent, T. N.) Арент, Т. Н.: Исследования по санитарной технике, ЛИСИ, Ленинград 1959.
15. (Medvedev, G. P.) Медведев, Г. П.: Санитарная техника **1**, 51 (1964).
16. (Levi, I. I.) Леви, И. И.: Моделирование гидравлических явлений, Госэнергоиздат, Москва 1960.
17. (Gordanov, T.) Горданов, Т.: Гидрология и Метеорология **6**, 3 (1965).
18. (Frankl, F. I.) Франкл, Ф. И.: О системе уравнений движения взвешенных наносов, Академий Наук СССР, Москва 1960.
19. (Mihailov, K. A.) Михайлов, К. А. и Богомолов, А. И.: Гидравлика, Госэнергоиздат, Москва 1964.
20. (Zrelov, M. P.) Зрелов, М. П.: Труды гидравлической лаборатории, Госэнергоиздат, Москва 1957.
21. Tesaker, E.: *Bull. River Harb. Res. Lab.* **15**, 169 (1976).
22. Bogárdi, J.: *Vízfolyások hordalékszállítása* (Sediment Transportation in Watercourses) Akadémiai Kiadó, Budapest 1971.

23. Bogárdi, J. and Szűcs, E.: *Acta Technica* **69**, 1 (1970).
24. Rouse, H.: *Proceedings, Hydraulics Conference, Bulletin* **20**, 33 (1940).
25. Yalin, S.: *Die Bautechnik* **36**, 3 (1959).
26. Yalin, S.: Similarity in sediment transport by currents. *Hydraulics Res. Paper* No. **6**, London 1965.
27. Komura, S.: *Trans. of JSCE* **80**,4 (1962).
28. Herbertson, J. G.: Scaling procedures for mobile bed hydraulic models in terms of similitude theory. JAHR Cong. **7**,3 (1969).
29. Zwamborn, J. A.: *La Houille Blanche* **3**, 5 (1966).
30. Einstein, H. A. and Ning Chien: *Trans. Am. Soc. Civil Engrs* **121**, 73 (1956).
31. Thompson, D. M.: Scaling laws for rectangular sedimentation tanks. Ph. D. Thesis. UMIST, U. K. (1967).
32. Thompson, D. M.: *Proc. Inst. of Civ. Engrs* **43**, 453 (1969).
33. Barr, D. I. H.: *Inst, Publ. Health Engrs J.* **63**, 175 (1964).
34. Humphreys, H. W.: *Journal AWWA* **67**, 7 (1975).
35. Clements, M. S.: *Proc. Inst. of Civ. Engrs* **34**, 171 (1966).
36. Clements, M. S. and Khattab, A. F. M.: *Proc. Inst. of Civ. Engrs* **40**, 471 (1968).
37. Villemonte, J. R. and Rohlich, G. A.: *Proc. of Purdue Univ.* 682 (1969).
38. Albrecht, A. E.: Investigation of similarity laws for circular sedimentation basins. M.Sc. Thesis, University of Wisconsin (1961).
39. Dague, R. R. and Baumann, E. R.: Hydraulics of circular settling tanks determined by models. Iowa Water Poll. Control Assoc. 1961.
40. Burdych, J.: *Vodni Hospodarstvi* **1**, 1 (1962).
41. Tessendorf, H.: *Das Gas- und Wasserfach* **115**, 349 (1974).
42. Christie, I. F. and Harbinson, R. W.: *Proc. Inst. of Civ. Engrs* **65**, 71 (1978).
43. Hubbel, G. E.: *Journal AWWA* **30**, 335 (1938).
44. Porteus, G. C.: The proportioning of rectangular sewage settlement tanks. M. Sc. Thesis. University of London (1964).
45. White, J. B. and Greenless, J. H : Model and prototype flow tests in circular sedimentation tanks. Symposium on present and future research on hydraulics of sewerage and sewage disposal. Introductory note 8, Inst. of Civ. Engrs, London 1962.
46. Vágás, I.: *Hidr. Közl.* **35**, 327 (1955).
47. Öllős, G.: *Hidr. Közl.* **4**, 253 (1958).
48. (Shifrin, S. M.) Шифрин, С. М.: Госиздательство литератур по строительству и архитектуре, Ленинград—Москва 1956.
49. Burdych, J.: (Interim Publication) Praha–Podbaba 1964.
50. Szalay, M.: *Hidr. Közl.* **36**, 141 (1960).
51. Fischerström, H. C.: *Proc. ASCE* Sept. (1955).
52. Eckenfelder, W. W. and O'Connor, D. J.: *Biological Waste Treatment*, Pergamon Press, New York 1961.
53. Conway, R. A. and Edwards, V. H.: *Chem. Engrg.* **6**, 167 (1961).
54. Camp, T. R.: *Sewage Works Journal* **7**, 742 (1936).
55. Schmidt-Bregas, F.: *Veröffentlichungen des Inst. für Siedlungswasserwirtschaft der TH. Hannover* **3**, (1958).
56. Groche, D.: *Stuttgarter Berichte zur Siedlungswasserwirtschaft*. Kommissionsverlag R. Oldenburg, München 1964.
57. Horváth, I.: *Hidrológiai Tájékoztató* 38, (1972).
58. Li, Kun: Research Center, Jones and Laughlin Steel Corp. Pittsburg, Pa. 1958.

59. Bramer, H.C. and Hoak, R. D.: *Ind. Eng. Chem. Process Design Develop.* **1**, 3, 185 (1962).
60. Rohlich, G. A.: *Proc. Am. Petrol. Inst.* **31**, 63 (1951).
61. Bramer, H. C. and Hoak, R. D.: *Ind. Eng. Chem. Process Design Develop.* **3**, 46 (1964).
62. Horváth, I.: Szakvélemény. Szerves Vegyipari Kutató Intézet (Expert's report. Research Institute for Organic Chemistry) Budapest 1970.
63. Horváth, I.: Mechanikai szennyvíztisztítás. (The primary stage of wastewater treatment) VMGT 50, VIZDOK, Budapest 1973.
64. (Velikanov, M. A.) Великанов, М. А.: Известия НИИГ, Ленинград 1934.
65. (Velikanov, M. A.) Великанов, М. А.: Динамика русловых потоков, Гидрометеориздат, Ленинград 1946.
66. Mau, G. E.: *Sew. Ind. Wastes* **31**, 1349 (1959).
67. (Pikalov, F. I.) Пикалов, Ф. И.: Гидротехника и Мелиорация, Сельхозгиз, Москва 1952.
68. Ivicsics, L.: *Hidromechanikai modellkísérletek*. (Hydraulic models) Műszaki Könyvkiadó, Budapest 1968.
69. Ivicsics, L.: *Hidr. Közl.* **3**, 75 (1957).
70. Camp, T. R.: *Transactions ASCE* **111**, 895 (1946).
71. Eliassen, R.: *Transactions ASCE* **111**, 947 (1946).
72. Horváth, I.: *Vízügyi Közlemények*, **4**, 527 (1980).
73. Kovács, Gy.: *A szivárgás hidraulikája* (The Hydraulics of Seepage) Akadémiai Kiadó, Budapest 1972.
74. Horváth, I.: *Österreichische Wasserwirtschaft* **5/6**, 129 (1965).
75. Mosonyi, E. and Kovács, Gy.: *Hidr. Közl.* **7/8**, 274 (1952).
76. Horváth I.: Szűrés a víz- és a szennyvíztechnológiában. BME Továbbképző Intézet kiadványa (Filtering in water and wastewater technology. Publication of the Institute of Extension Courses of the Technical University of Budapest) Budapest 1972.
77. Horváth, I.: *Hidr. Közl.* **3**, 217 (1962).
78. Ison, C. R. and Ives, K. J.: *Chem. Engrg. Sci.* **24**, 717 (1969).
79. Ison, C. R.: *Ph.D. Thesis*. University of London (1967).
80. Ives, K. J.: *Water Research* **4**, 201 (1970).
81. Ives, K. J.: Theory of filtration. Int. Water Supply Congress, Vienna 1969.
82. Ives, K. J.: *Effluent Water Treatment Journal* **6**, 522, 591 (1966).
83. Fitzpatrick, J. A. and Spielman, L. A.: *Journal of Colloid and Interface Science* **43**, 350 (1973).
84. Spielman, L. A. and Fitzpatrick, J. A.: *Journal of Colloid and Interface Science* **42**, 607 (1973).
85. Donovan, E. J.: *High-rate Filtration of Industrial Wastes*. McGraw-Hill, New York 1972.
86. Adin, A., Baumann, E. R. and Cleasby, J. L.: *Journal AWWA* **1**, 17 (1979).
87. Juhász, J.: Mélységi szűrők vizsgálata. VITUKI zárójelentés (Investigation of deep-bed filters. Final report of VITUKI) Budapest 1974.
88. Zingler, E.: Die Filtration von Abwasser schlämmen I—II. *Veröffentlichungen des Instituts für Stadtbauwesen* TH. Braunschweig 1969–1970.
89. Sennett, P. and Olivier, J. P.: *Ind. and Engrg. Chem.* **57**, 32 (1965).
90. Camp, T. R. and Stein, P. C.: *J. Boston Soc. Civ. Engrg.* **30**, 219 (1943).
91. Rufy, J., Hudson, H. E. and Singley, J. E.: *Journal AWWA* **67**, 10 (1975).

92. Camp, T. R.: *Trans. Am. Soc. Civ. Engrs* **120,** 1 (1951).
93. Souček, J. and Sindelar, J.: *The Use of a Dimensionless Criterion in the Characterization of Flocculation.* Praha–Podbaba 1967.
94. Ives, K. J.: *Proc. Inst. of Civ. Engrs* **39,** 243 (1968).
95. Delichatsios, M. A. and Probstein, R. F.: *Journal WPCF* **45,** 941 (1975).
96. Varrók, E.: Az Ajkai Hőerőmű vízlágyító berendezéseinek áramlástani vizsgálata. VITUKI jelentés (Hydraulic investigation of the water softening equipment at Ajka. Final report of VITUKI) Budapest 1963.
97. Ivicsics L.: VITUKI beszámoló, 1966 (Report of VITUKI, 1966) Budapest 1971.
98. Gould, B. W.: *Vodohosodársky časopis.* **4,** 459 (1968).
99. Gould, B. W.: *The Civil Engineering Trans.* **4,** 55 (1969).
100. Gould, B. W.: *Effl. Water Treatment Jour.* **14,** 457 (1974).
101. Tesařik, I.: *Wasserwirtschaft–Wassertechnik* **9,** 172 (1959).
102. Tesařik, I.: *Wasserwirtschaft–Wassertechnik* **11,** 189 (1961).
103. Tesařik, I.: Discussion. *Proc. Inst. of Civil Engrs* **40,** 569 (1968).
104. Ives, K. J.: *Proc. Inst. of Civ. Engrs* **39,** 243 (1968).
105. (Pervov, G. G.) Первов, Г. Г.: Труды Института ВОДГЕО, **3,** 53 (1963).
106. Water Research Association, Marlow, England. 10th Annual Report, Dec. 1964.
107. Letterman, R. D., Quon, J. E. and Gemmell, R. S.: *Journal WPCF* **46,** 2536 (1974).
107a. Sherwood, T. K. and Brian, P. L. T.: *Research and Development.* **Progress Report,** No. 334 (1968).
107b. Calderbank, P. H. and Moo-Young, M. B.: *Chem. Eng. Sci.* **16,** 39 (1961).
108. Sherwood, T. K.: *Trans. Amer. Inst. Chem. Eng.* **36,** 817 (1940).
109. Frössling, N.: *Gerlands Beitr. Geophys.* **52,** 170 (1938).
110. Cornwell, D. A. and Zoltek, J.: *Journal WPCF* **4,** 600 (1977).
111. Ryon, A. D., Daley, F. L. and Lowrie, R. S.: Design and scale-up of mixer settlers for the Dapex solvent extraction process. USAEC ORNL 2951 (1960).
112. Treybal, R. E.: *Liquid extraction.* McGraw-Hill, New York 1963.
113. Milbury, W. F. and Stack, V. T.: A laboratory and full-scale study of the effect of mixing configuration on phosphorus removal in the activated sludge system. 26th Industrial Waste Conf., Purdue University, Lafayette 1971.
114. Drnevich, R. F.: Biological-chemical process for removing phosphorus at Reno/Sparks, N. V. U. S. EPA., Grant No. 804931, Cincinnati, OH. 1979.
115. Hart, F. L. and Gupta, S. K.: *Journal of the Environmental Engrg. Div., Proc. ASCE* **EE4,** 785 (1978).
116. Sawyer, C. M. and King, P. H.: The hydraulic performance of chlorine contact tanks. Proceedings, 24th Industrial Waste Conf., Purdue University, Lafayette 1969.
117. Louie, D. S. and Fohrman, M. S.: *Journal WPCF* **40,** 174 (1968).
118. Mailer, W. J., Behn, V. C. and Gates, Ch. D.: *Journal of the San. Engrg. Div., Proc. ASCE* **SA4,** 91 (1967).
119. Swilley, E. L. and Atkinson, B.: Proceedings, 8th Industrial Conf., Purdue University, Lafayette 706 (1963).
120. Quirk, Th. P.: Scale-up and process design techniques for fixed film biological reactors. Conference of Int. Assoc. for Water Pollution Res. Jerusalem 1972.
121. Kehrberger, G. J.: Effect of recirculation on the performance of a trickling filter. *Ph. D. Thesis.* Rice University, Houston (1968).

122. Sinkoff, M. D., Porges, R and McDermott, J. H.: *Journal of the San. Engrg. Div., Proc. ASCE* **85**, SA6, 662 (1959).
123. Gerber, B.: *Sewage and Ind. Wastes* **26**, 136 (1964).
124. (Yakovlev, S. V. and Galanin, P. I.) Яковлев, С. В. and Галанин, П. И.: Водоснабжение и Санитарная Техника, **6**, 4 (1957).
125. Tuček, F. and Chudoba, J.: *Sborník Vysoké školy chem.-technol. Praze.* **F11**, 5 (1966).
126. Tuček, F., Chudoba, J. and Madera, V.: *Water Research* **5**, 647 (1971).
127. Hartmann, L.: *Journal WPCF* **39**, 958 (1967).
128. Williamson, K. and McCarthy, P. L.: *Journal WPCF* **48**, 9 (1976).
129. Eckenfelder, W. W. and Cardenas, R. R.: *Biotechnology and Bioengineering* **8**, 389 (1966).
130. Jones, K.: *Journal Inst. Sew. Purif.* **5**, 478 (1965).
131. Fischerström, N. C. H.: *Journal of the San. Engrg. Div., Proc. ASCE* **86**, SA5, 21 (1960).
132. Robertson, W. S.: *Journal and Proceedings. The Institute of Sewage Purif.* **6**, 585 (1964).
133. Knop, E. and Kalbskopf, K. H.: *Das Gas- und Wasserfach* **110**, 226 (1969).
134. Kalbskopf, K. H.: Strömungsverhältnisse in Belüftungsbecken mit Kreiselbelüftern (Manuscript) 1971.
135. Kemény, I.: *Hidr. Közl.* **8**, 358 (1975).
136. Kalinske, A. A., Shell, G. L. and Lash, L. D.: *Water and Wastes Engineering* **4**, 65 (1968).
137. Zeper, J. and De Man, A.: New developments in the design of activated sludge tanks with low B. O. D. loadings. Int. Conf. on Water Pollut. Res., San Francisco (1970).
138. Lewandowski, B., Borowski, J. and Rembeza, L.: *Gospodarka Wodna* **3**, 84 (1969).
139. Harremoës, P.: Dimensionless analysis of circulation, mixing and oxygenation in aeration basin. IAWPR-Specialized Conference on Aeration, Amsterdam, 1978.
140. Maise, G.: *Journal of the San. Engrg. Div., Proc. ASCE* **96**, SA5, 1079 (1970).
141. Thomas, H. A. and McKee, J. E.: *Sewage Works Journal* **16**, 42 (1944).
142. McLean, H.: *Ph. D. Thesis.* McMaster University, Hamilton, Ontario (1966).
143. Murphy, K. L. and Timpany, P. L.: *Journal of the San. Eng. Div., Proc. ASCE* **93**, SA5, 1 (1967).
144. Murphy, K. L. and Boyko, B. I.: *Journal of the San. Eng. Div., Proc. ASCE* **96**, SA2, 211 (1970).
145. Hartmann, L. and Laubenberger, G.: *Journal WPCF* **40**, 670 (1968).
146. Laubenberger, G.: Struktur und physikalisches Verhalten der Belebtschlammflocke. *Karlsruher Berichte* **3** (1970).
147. Kalinske, A. A.: Turbulence diffusivity in activated sludge aeration basins. 5th Int. Conf. on Water Pollut. Res., San Francisco 1970.
148. Fair, G. M., Gemmel, R. S. and Myrick, H. N.: *Advances in Water Pollut. Res.*, Proc. 2nd Int. Conf. Water Pollut. Res., Pergamon Press, **2**, 201 (1965).
149. Zahradka, V.: The role of aeration in the activated sludge process. 3rd Int. Conf. on Water Pollut. Res., Munich 1968.
150. Rushton, J. H., Costich, E. W. and Everett, H. J.: *Chem. Engrg. Progress* **46**, 395 (1950).
151. Rushton, J. H.: *Chem. Engrg. Progress* **47**, 485 (1951).

152. Rushton, J. H. and Oldshue, J. Y.: *Chem. Engrg. Progress* **49**, 161 (1953).
153. Oldshue, J. Y.: *Biological Treatment of Sewage and Industrial Wastes.* Vol. I, 231., Reinhold Publishing Corporation, New York 1956.
154. Oldshue, J. Y.: *Journal of Paint Technology* **40**, 66 (1968).
155. Connolly, J. R. and Winter, R. L.: *Chem. Engrg. Progress* **65**, 70 (1969).
156. Aiba, S., Humphrey, A. E. and Millis, N. F.: *Biochemical Engineering.* University of Tokyo 1965.
157. Biggs, R. D.: *Am. Inst. Chem. Eng. J.* **9**, 636 (1963).
158. Fox, E. A. and Gex, V. E.: *Am. Inst. Chem. Eng. J.* **2**, 539 (1956).
159. Norwood, K. W. and Metzner, A. B.: *Am. Inst. Chem. Eng. J.* **6**, 432 (1960).
160. Hansford, G. S. and Humphrey, A. E.: *Biotechnology and Bioengineering* **8**, 85 (1966).
161. Boyle, W. C. (Ed.): Proceedings, Workshop toward an oxygen transfer standard. EPA-600/9-78-021, 1979.
162. Kalinske, A. A.: Scale-up problems relating to aeration activated sludge in waste treatment. Microbial Chemistry and Technology. Div. of Am. Chem. Soc. 150th National Meeting, Salt Lake City 1965.
163. Danckwerts, P. V.: *Journal AJChE* **I**, 4, 456 (1955).
164. Kiiskinen, S.: Comparison of different aerators for diffused aeration. IAWPR-Specialized Conference on Aeration, Amsterdam 1978.
165. Kataoka, H. and Miyauchi, R.: *Kagaku Kogaku* **30**, 409 (1966).
166. Oldshue, J. Y.: *Adv. Appl. Microbiol.* **2**, 275 (1960).
167. Jackson, M. L. and Collins, W. D.: *Ind. and Engrg. Chem.* **3**, 386 (1964).
168. Horváth, I.: *Építés- és Közlekedéstudományi Közlemények* **3–4**, 551 (1964).
169. Horváth, I.: *MTA VI. Osztály Közleményei,* **36**, 105 (1965).
170. Quigley, J. T. and Boyle, W. C.: *Journal WPCF* **48**, 357 (1976).
171. Schmidtke, N. W.: Scale-up methodology of surface aerated model reactors. Ph. D. Thesis. University of Waterloo (1974).
172. Schmidtke, N. W. and Horváth, I.: Scale-up methodology for surface aerated reactors. Int. Conf. on Water Pollut. Res., Sydney 1976.
173. Horváth, I.: Modelling of oxygen transfer processes in aeration tanks. 3rd Int. Conf. on Water Pollut. Res., Munich, WPCF, Washington 1966.
174. Kalinske, A. A.: *Journal of the San. Engrg. Div., Proc. ASCE* **94**, SA3, 575 (1968).
175. Cleasby, J. L. and Baumann, E.: *Journal WPCF* **40**, 412 (1968).
176. von der Emde, W.: Belüftungsysteme und Beckenformen. *Vom Wasser.* Wien 1970.
177. Kalinske, A. A.: Biological Treatment of Sewage and Industrial Wastes. Vol. I, 241, Reinhold Publishing Corporation, New York 1956.
178. Rincke, G.: Discussion. 3rd Int. Conf. on Water Pollut. Res., Munich 1966.
179. Zlokarnik, M.: Sorption characteristics for gas-liquid contacting in mixing vessels. *Advances in Biochemical Engineering 8.* Springer-Verlag, Berlin-Heidelberg 1978.
180. Zlokarnik, M.: Scale-up of surface aerators for wastewater treatment. *Advances in Biochemical Engineering 11.* Springer-Verlag, Berlin-Heidelberg 1979.
181. Groot Wassink, J., Rácz, I. G. and Goinga, C. R.: Process study of the behaviour of a new type of mechanical aerator in square aeration tanks. Int. Symposium on Mixing. Paper C10, Mons, Belgium 1978.
182. Horváth, I.: *Hidr. Közl.* **11**, 494 (1966).
183. Horváth, I.: Levegőztető medencék hidraulikai és oxigénfelvételi folyamatainak mo-

dellvizsgálata. VITUKI kiadvány (Model studies on the hydraulics and oxygen transfer processes in aeration tanks. Publication of VITUKI) No. **29,** Budapest 1970.
184. Roustan, M.: Performance des aerateurs de surface. Int. Symposium on Mixing. Paper C9, Mons, Belgium 1978.
185. Bruxelmane, M.: Comportement hydrodinamique et efficacite de transfert des turbines a disque utilisses comme aerateurs de surface. Int. Symposium on Mixing. Paper C11, Mons, Belgium 1978.
186. Grassmann, P.: *Physikalische Grundlagen der Chemie-Ingenieur-Technik.* Verlag Sauerländer, Aarau 1961.
187. Levich, V. G.: *Fizikai-kémiai hidrodinamika* (Physico-chemical hydrodynamics) Akadémiai Kiadó, Budapest 1958.
188. Horváth, I.: Levegőztető rendszerek a szennyvíztechnológiában. BME Továbbképző Int. kiadványa (Aerating systems in wastewater technology. Publication of the Institute of Extension Courses of the Technical University of Budapest) 1975.
189. Cooper, C. M., Fernstrom, G. A. and Miller, S. A.: *Ind. and Engrg. Chem.* **36,** 504 (1944).
190. Karow, E. O., Bartholomew, W. H. and Sfat, M. R.: *Journal Agr. Food Chem.* **1,** 302 (1953).
191. Bartholomew, W. H.: *Adv. Appl. Microbiol.* **2,** 289 (1960).
192. Finn, R. K.: *Bacteriol. Revs.* **18,** 254 (1954).
193. Hixon, A. W. and Gaden, E. L.: *Ind. and Engrg. Chem.* **42,** 1792 (1950).
194. Oldshue, J. Y.: *Biotechnology and Bioengineering* **8,** 3 (1966).
195. Sherwood, T. K. and Pigford, R. L.: *Absorption and extraction.* McGraw-Hill, New York 1952.
196. Strom, J., Dale, H. F. and Peppler, H. J.: *Appl. Microbiol.* **7,** 235 (1959).
197. Wegrich, O. G. and Shurter, R. A.: *Ind. and Engrg. Chem.* **45,** 1153 (1953).
198. Jensen, A. L., Schulz, J. S. and Shu, P.: *Biotechnology and Bioengineering* **8,** 525 (1966).
199. Madera, V., Tuček, F. and Chudoba, J.: *Sborník Vysoké školy chem.-technol. Praze.* F **12,** 77 (1967).
200. Sundstrom, D. W., Klei, H. E. and Molvar, A. E.: *Water Research* **7,** 1905 (1973).
201. Horváth, I.: *Das Gas- und Wasserfach* **12,** 311 (1967).
202. Horváth, I.: Derzeitiger Stand des Problems der Ähnlichkeit und Maßvergrößerung auf dem Gebiet der Abwasserreinigung. VITUKI-Bericht No. **5,** Budapest 1969.
203. Boyle, W. C. and Rohlich, G. A.: *Biotechnology and Bioengineering* **8,** 405 (1966).
203a. Pasveer, A.: *Sewage and Industrial Wastes* **26,** 28 (1954), **27,** 1130 (1955), **28,** 28 (1956).
204. Eckenfelder, W. W., Goodman, B. L. and Englande, A. J.: *Advances in Biochemical Engineering 2,* Springer-Verlag, Berlin–Heidelberg–New York 1972.
205. Mancini, J. L. and Barnhart, E. L.: *Adv. Water Quality Improv.* **1,** 79 (1968).
206. Quirk, T. P.: *Water and Wastes Engrg.* **6,** D-1 (1969).
207. Gates, W. E., Marlar, J. T. and Westfield, J. D.: *Water Research* **3,** 663 (1969).
208. Ludžack, F. J.: *Journal WPCF* **32,** 605 (1960).
209. Ford, D. L.: Proc. of Seminar, Nashville, Tenn. Vanderbilt University 1970.
210. Staff of the Water Pollut. Res. Laboratory: Control of the flow of liquids in small-scale plant. *J. Inst. Sew. Purif.* **4,** 393 (1951).
211. Naito, M., Takamatsu, T., Fan, L. T. and Lee, E. S.: *Biotechnology and Bioengineering* **11,** 731 (1969).

212. Burkhead, C. E. and Wood, D. J.: *Journal of the San. Eng. Div., Proc. ASCE* SA3, 593 (1969).
213. Silveston, P. L.: Int. Conf. on Water Pollut. Res., Jerusalem 1972.
214. Erickson, L. E., Ho, Y. S. and Fan, L. T.: *Journal WPCF* **40**, 717 (1968).
215. Tebbutt, T. H. Y. and Christoulas, D. G.: *Wat. Pollut. Control* 701 (1975).
216. Bhatla, M. N., Stack, V. T. and Weston, R. F.: *Journal WPCF* **38**, 601 (1966).
217. Vath, C. A., Sitman, W. D. and Weston, R. F.: *The significance of batch treatment kinetics in activated sludge design.* Weston Inc., Publication, West Chester 1972.
218. Blanch, H. W. and Dunn, I. J.: Modelling and simulation in biochemical engineering. *Advances in Biochemical Engineering 3,* Springer Verlag, Berlin–Heidelberg–New York 1974.
219. Chesner, W. H. and Molof, A. H.: *Prog. Wat. Tech.* **9**, 811 (1977).
220. Ouano, E. A. R.: *Prog. Wat. Tech.* **9**, D 18 (1977).
221. Ouano, E. A. R.: *Water Research* **12**, 1005 (1978).
222. Bintanja, H. H. J., Van der Erve, J. J. V. M. and Boelhouwer, C.: *Water Research* **9**, 1147 (1975).
223. Deindoerfer, F. H. and West, J. M.: *Adv. Appl. Microbiol.* **2**, 265 (1960).
224. Erdmenger, R.: *Chemie-Ing. Techn.* **36**, 175 (1964).
225. Ward, P.: *Journal Inst. Sew. Purif.* **1**, 85 (1955).
226. Bayley, R. W.: *Proc. Soc. Water Treatment and Exam.* **10**, 26 (1961).
227. Torpey, W. N.: *Sewage and Industrial Wastes* **27**, 121 (1955).
228. Schlenz, H. E.: *Sewage and Industrial Wastes* **27**, 133 (1955).
229. Heukelekian, H.: *Sewage and Industrial Wastes* **27**, 142 (1955).
230. Dick, R. J. and Ewing, B. B.: *Journal WPCF* **39**, 543 (1967).
231. Wiedemann, F.: *Das Gas- und Wasserfach* **118**, 1 (1977).
232. Pallasch, O. and Triebel, W.: *Lehr- und Handbuch der Abwassertechnik.* Band I–III. Verlag von Wilhelm Ernst und Sohn, Berlin–München 1967.
233. Horváth, I.: *Hidr. Közl.* **2**, 56 (1980).
234. Horváth, I.: *Das Gas- und Wasserfach* **7**, 317 (1981).
235. Horváth, I.: *Hidr. Közl.* **3**, 116 (1978).
236. Hibberd, R. L.: *Wat. Pollut. Control* **5**, 14, (1974).
237. Edde, H. J.: *Ph. D. Thesis.* University of Texas (1967).
238. Edde, H. J. and Eckenfelder, W. W.: *Journal WPCF* **40**, 1486 (1968).
239. Stalmann, V.: *Veröffentlichungen des Institutes für Siedlungswasserwirtschaft der TH. Hannover* (1965).
240. Eckenfelder, W. W. and Ford, D. L.: *Water pollution control experimental procedures for process design.* Jenkins Book Publ. Austin, Texas 1970.
241. Horváth, I., Nagy, L. and Bánvölgyi, I.: Találmány. (Invention) No. 163 651 (1971).
242. Horváth, I., Nagy, L. and Bánvölgyi, I.: Találmány. NSZK szabadalom. (Invention. GFR Patent.) No. 2. 254, 171 (1974); Svéd szabadalom (Swedish Patent) No. 72. 14057.7 (1975).
243. Horváth, I.: *Hidr. Közl.* **3**, 134 (1965).
244. Horváth, I.: A hatvani és az angyalföldi szellőztető-medencék kismintavizsgálata. Beszámoló a VITUKI 1962. évi munkájáról (Model studies on aeration tanks in Hatvan and Angyalföld. Report of the work of VITUKI in 1962) Budapest 1965.
245. Horváth, I.: *Hidr. Közl.* **3**, 257 (1963).
246. Horváth, I.: *Hidr. Közl.* **7**, 307 (1970).
247. Horváth, I.: *Research in Water Quality and Water Technology* **1**, 79 (1972).

248. Horváth, I.: Levegőztető medencék oxigénfelvételi folyamatainak kismintavizsgálata. (Modelling of oxygen transfer processes in aeration tanks.) *C. Sc. Thesis.* Hungarian Academy of Sciences, Budapest (1971).
249. Horváth, I.: *MTA VI. Osztály Közl.* **38,** 125 (1967).
250. Horváth, I.: *Hidr. Közl.* **3,** 127 (1969).
251. Horváth, I.: *Magyar Kémikusok Lapja* **8,** 384 (1970); **10,** 517 (1970).
252. Horváth, I.: *Magyar Kémikusok Lapja* **1,** 8 (1972).
253. Horváth, I.: *Hidr. Közl.* **10,** 436 (1972).
254. Horváth, I.: *Műszaki Tudomány* **46,** 71 (1973).
255. Herbert, D.: *Chem. Soc. Ind. Monograph* **12,** 21 (1961).
256. Hinshelwood, C. N.: *The Chemical Kinetics of the Bacterial Cell.* Oxford University Press, London 1946.
256a. Märki, E.: *Gas-Wasser-Abwasser* **49,** 13 (1969).
257. Dobbins, W. E.: Discussion. 3rd Int. Conf. on Water Pollut. Res., Munich, WPCF, Pergamon Press, Washington 1966.
258. Rousse, G. and Brouzes, P.: Discussion. 3rd Int. Conf. on Water Pollut. Res., Munich, WPCF, Pergamon Press, Washington 1966.
259. Inoue, Y.: Discussion. 3rd Int. Conf. on Water Pollut. Res., Munich, WPCF, Pergamon Press, Washington 1966.
260. Möller, U.: *Das Gas- und Wasserfach* **108,** 311 (1967).
261. Wilderer, P. and Hartmann, L.: *Das Gas- und Wasserfach* **110,** 707 (1969).
262. Hammerton, D. and Gamer, F. H.: *Trans. Inst. Chem. Eng.* **32,** Supplement (1954).
263. Horváth, I.: *Das Gas- und Wasserfach* **106,** 1343 (1965).
264. Kalinske, A. A.: *Proc. ASCE* SA3, 575 (1968).
265. Rácz, I. G.: *BECEWA* **45,** 27 (1978).
266. Horváth, I.: Some questions of the scale-up aeration systems. IAWPR-Specialised Conference on Aeration. The Netherlands National Committee of the IAWPR on Water Pollution Research, Amsterdam 1978.
267. Horváth, I.: *Progress in Water Technology* **11,** 73 (1979).
268. Clark, E.: Discussion. IAWPR-Specialised Conference on Aeration. The Netherlands National Committee of the IAWPR on Water Pollution Research, Amsterdam 1978.
269. Salamin, P.: *Az ÉKME Tudományos Közleményei* **3–4,** 337 (1965).
270. Horváth, I.: *Hidr. Közl.* **6,** 280 (1964).
271. Benedek, P.: *Hidr. Közl.* **6,** 282 (1964).
272. Hock, B.: *Hidr. Közl.* **6,** 282 (1964).
273. Dickerson, R. G.: The economic aspects of scaling-up chemical plants. Joint Symposium on Scaling-up, S15, Inst. Chem. Engrs, London 1957.
274. Walley, K. H. and Robinson, S. J. Q.: *Magyar Kémikusok Lapja* **7,** 360 (1973).
275. Tchobanoglous, G.: *Public Works* **7,** 61 (1974), **8,** 58 (1974).
276. Rachford, T. M., **Scarato,** R. F. and Tchobanoglous, G.: *Journal of the San. Eng. Div., Proc. ASCE* **95,** SA6, 1063 (1969).
276a. Ábrahám, O. and Takács, A.: *Hidr. Közl.* **11,** 481 (1972).
277. Negaard, J.: *Gas-Wasser-Abwasser* **56,** 18 (1976).
278. Tihansky, D. P.: *Journal WPCF* **46,** 813 (1974).
279. Heberling, G. and Hahn, H. H.: *Das Gas- und Wasserfach* **116,** 206 (1975).
280. Young, C. E. and Carlson, G. A.: *Journal WPCF* **47,** 2563 (1975).
281. Horváth, I.: Válogatott fejezetek a szennyvíztisztítás és a vízelőkészítés köréből. BME

Továbbképző Intézet kiadványa (Selected chapters in the field of water and wastewater treatment. Publication of the Institute of Extension Courses of the Technical University of Budapest) Budapest 1970.
282. Horváth, I.: Levegőztető rendszerek a szennyvíztechnológiában. BME Továbbképző Intézet kiadványa (Aeration systems in the wastewater technology. Publication of the Institute of Extension Courses of the Technical University of Budapest) Budapest 1975.
283. Horváth, I.: *Das Gas- und Wasserfach* **115,** 128 (1974).
284. Horváth, I.: A csatornázás és szennyvízkezelés költségszámítása (Cost estimation for sewerage and wastewater treatment) VMGT 88, VIZDOK, Budapest 1977.
285. Horváth, I.: Rendszertechnika és optimalizálás a csatornázás és a szennyvízkezelés területén (System technics and optimalization in the field of hydraulics of sewerage and wastewater treatment) VMGT 89, VIZDOK, Budapest 1977.
286. Horváth, I.: A csatornázás és a szennyvízkezelés hidraulikája. (Hydraulics of sewerage and wastewater treatment) VIZDOK, Budapest 1976.
287. Horváth, I.: A csatornázás és szennyvízkezelés hidraulikája. Példatár (Hydraulics of sewerage and wastewater treatment. Exercises) VIZDOK, Budapest 1979.

SUMMARIZING BOOKS AND PAPERS

288. Aiba, S., Humphrey, A. E. and Millis, N. F.: *Biochemical Engineering*. University of Tokyo, Tokyo 1965.
289. Benedek, P. and László, A.: *A vegyészmérnöki tudomány alapjai* (The bases of chemical engineering science) Műszaki Könyvkiadó, Budapest 1964.
290. Fejes, G.: *Ipari keverőberendezések* (Industrial mixing equipment) Műszaki Könyvkiadó, Budapest 1970.
291. Horváth, I.: A csatornázás és szennyvízkezelés hidraulikája. (Hydraulics of sewerage and wastewater treatment) VIZDOK, Budapest 1976.
292. Horváth, I.: A modellalkotás mint tudományos kutatási módszer (Model-creation as a scientific research method) *Magyar filozófiai szemle*, **2,** 161 (1965).
293. Johnstone, R. E. and Thring, M. W.: *Pilot Plants, Models and Scale-Up Methods in Chemical Engineering*. McGraw-Hill, New York 1957.
294. (Kirpichev, M. V. and Konakov, P. K.) Кирпичев, М. В. and Конаков, П. К.: Математические основы теории подобия. АН СССР, Москва 1949.
295. Klinkenberg, A. and Moody, H. H.: *Chem. Engrg. Progress* **44,** 1, 17 (1948).
296. Langhaar, H. L.: *Dimensional Analysis and Theory of Models*. John Wiley, New York 1951.
297. Levenspiel, O.: *Chemical Reaction Engineering*. John Wiley, New York 1962.
298. Purchas, D. B. (Ed.): *Solid/Liquid Separation Equipment Scale-Up*. Uplands Press, London 1978.

PAPERS CONTAINING COLLECTION OF DIMENSIONLESS NUMBERS

299. Boucher, D. F. and Alves, G. E.: Dimensionless numbers. 1–2. *Chem. Engrg. Progress* **55**, 9, 55 (1959); **59**, 8, 75 (1963).
300. Catchpole, J. P. and Fulford, G. D.: Dimensionless groups. *Ind. and Engrg. Chem.* **58**, 3, 46 (1966).
301. Fulford, G. D. and Catchpole, J. P.: Dimensionless groups. *Ind. and Engrg. Chem.* **60**, 3, 71 (1968).

Subject index

Absorption 113, 114
Activated sludge 89
   system 112, 122, 125, 126, 163, 167
   treatment 69, 148
Adhesion number 49
Adsorption, activated carbon 66
Aeration tank 78
Air-lift pump 139, 141
Analogue simulation 131
Approximate method 6
Archimedean number 15, 64
Attenuation curve 41
Axial dispersion 86

Bacterial population 129
Band filter 51
Biological treatment 74, 106, 131
Bodenstein number 86, 87
BOD loading 166
Boltzmann's constant 49
Boundary conditions 51, 123, 129, 169, 170
Bubble aeration 122
Bubble movement 121

Cake filtration 51
Camp number 48, 53–57, 59
Capillary (Horváth) number 47
Carrousel-type aeration system 81
Chlorination chamber 71
Circular settling basin 23
Clarifier 60
   horizontal-flow 57
   similarity of upward-flow 63

Coagulation 52, 59
   orthokinetic 56
   perikinetic 56
Coalescence index 113
Complete mixing 127
Compressed-air aeration system 101, 149
Contact chamber 74
Cost function 177, 179, 180

Damköhler number 158, 165
Damköhler's equation 158, 159, 160
Deflocculation 57, 59
Density current 22
Density ratio 14, 15
Digester 133
Digestion index 137
Dilution rate 98
Dimensional analysis 11, 16, 20, 33, 47, 67, 75, 76, 83, 88, 90, 102, 113, 115, 163
Discharge coefficient 105
Disinfection 70
Dispersion 86
Dissimilar fermenter 126
Dissimilar systems 139, 168
Distorted model 95
Distortion 141
Dobbins–Camp theory 21
Dortmund tank 27, 28, 29, 71, 74

Economic similarity 144, 175, 182
Electrophoretic mobility 51, 52
Elutriation 69, 70
Endogenous respiration 127
Energy dissipation 52, 54, 60, 67, 92

Energy per unit volume 94, 95
Erdmenger's concept 135
Eulerian number 84, 93, 97, 104, 151, 163
Extraction 69
Extrapolation 180, 183

Fermentation, continuous 98
Fermentation process 122, 123, 133
Fermenter 122, 123, 136
  dissimilar 126
Fermenter scale-up 124
Filter 46
  band 51
  deep-bed 48
Filter press 51
Filtration 47
  cake 51
Flat-blade turbine 118, 120
Floating sludge blanket 60
Floating solid 8
Flocculating substances, sedimentation of 45
Flocculation 52, 54, 56, 57
Flow
  in porous media 47
  plug 127
  scale factor of air 12, 150, 158
  three-phase 47
  two-phase 18–20, 47
Flow rate control 130
Flow-through
  characteristics 71
  method 16, 23, 25, 27, 50, 71, 86, 87
Fluidized bed 66
Frictional resistance, coefficient of 54
Froude number 8, 9, 11, 12, 14–18, 21, 22, 21–33, 39, 43, 44, 46, 48, 60, 62–64, 70–75, 79–84, 93, 95, 97, 99–105, 107, 109, 115, 117–119, 121, 138, 149, 151, 163
Frössling equation 67, 68, 173
Full-scale study 6

Galilei number 15, 81, 116, 117, 119
Gas-fluid interface, renewal of 109
Generalization of scale-up method 31

Generalized Re number 134
Generation time 165
Geometrically dissimilar system 88, 109
Gilliland-Sherwood equation 67, 68, 173
Gordanov's approach 19
Gould's approach 61, 64
Gravitational number 49
Gravitational sludge thickener 145

Hamaker's constant 49
Hazen number 21, 25–27, 45, 46
Hazen's concept 21, 25, 43, 45
Head loss 7
Helmholtz-Smoluchowski expression 52
Homochronous number 59, 158, 159
Horváth number 47
Hydraulic efficiency 32

Identical biomass 129
Inertial parameter 48
Initial conditions 51, 123, 129, 169
INKA-type system 79, 148, 154, 155, 158, 159
Inlet ratio 23
Interception parameter 48
Invariant function 89, 154
Isotropic turbulence 59, 67

Kármán number 151, 152
Kessener-type system 148, 154, 155
Kirpitshov–Guhman's theorem 169
Kirschmer's expression 7
Kolmogorov's concept 59, 67
Koženy–Carman equation
Kynch's settling theory 62

Laboratory-scale experiments 129, 130
Levi's method 18
Longitudinal dispersion 88
Lyashchenko number 15

Macro-turbulence 91
Mass loading 146

Mass transfer 99, 101, 115, 123, 131, 160, 161
   coefficient, extended 100, 104, 107, 129
   number 103
Material number 115
Mathematical modelling 130
Method
   flow-through 16, 23, 25, 27, 50, 71, 86, 87
   generalization of scale-up 31
Michaelis constant 77, 127
Micro-turbulence 91, 92
Mixing 67, 93
   longitudinal 86, 88
Mixing chamber 74
Mixing time 97, 98
Model
   Monod's kinetic 130
   open-flow hydraulic 16
   Ostwald-de Waele 134
   undistorted 43
Modelling
   mathematical 130
   of mixing devices 96
   partial 124, 186
   trivial 37, 77, 164, 165, 167, 171
Model-prototype relationship 5, 27, 38, 43, 62, 120, 163
Model study 5, 6, 186
Monod's kinetic model 130
Mosonyi–Kovács number 11, 47,

Newton number 118, 119
Newton's law 14, 15
Non-Newtonian media 133, 134, 135, 136
Number involving zeta-potential 49
Nusselt number 131, 160

Oil separator 34
Ostwald-de Waele model 134
Oxygen absorption, efficiency of 111
Oxygen consumption 90
Oxygen deficiency 100, 103
Oxygen input 100, 102, 115
Oxygen transfer 110, 154, 156, 157, 162
Oxygenation 99
   economics of 119

Packing number 75
Paddle agitator 57
Peclet number 48, 49, 158, 160
Perfect similitude 186
Perikinetic coagulation 56
Phosphorus removal 69
Pilot-scale study 6, 129, 130
Pipe hydraulics 40
Plastic material 134
Plate separator 29, 30, 31, 44
Poiseuille number 47
Poisson's equation 52
Power consumption 162
Power correlation 116
Power input 117
Power number 84, 93, 97
Prandtl number 131
Primary stage treatment 7
Pseudoplastic material 134
Pulsation 152
Pumping capacity 96, 100

Reaction kinetics 163
Reaction number 75
Recirculation 76, 127, 167
Recirculation factor 168
Renewal of gas-fluid interface 109
Resolving the overall modelling problem 169
Retention time, dimensionless 127
Reynolds equations 152
Reynolds number 8, 11, 14–18, 21–23, 25–31, 33, 35, 40, 44, 46, 48, 57, 59, 60, 62–64, 67, 71, 75, 77, 81, 83, 84, 89, 91, 93–95, 97, 99–101, 103–105, 109, 115, 116, 119–121, 131–132, 134, 138, 151, 155–157, 163, 173
Richardson number 44, 61–63
Rotating disc biological equipment 131, 132
Roughness, similarity of 16
Rubey's relationship 14, 15

Sand trap 8
   aerated 9, 11
   Geiger-type 8
   tangential 8

Scale effect 6, 27, 35, 102, 142, 171, 183
Scale factor 3, 31, 179
Scale-up 5, 9, 23, 186
Scaling down 187
Scaling-up activated sludge system 128
Scaling-up digester 142
Scaling-up mixing equipment 98
Schmidt number 67, 94, 101, 104, 113, 115
Screen 7
Secondary current 29
Sedimentation index 33
Sediment transport 15, 20
Seepage 47
Self-modelling 16, 17, 19, 25, 28, 32, 58, 89, 109
Settling basin 13
  circular 23
  longitudinal-flow 16, 18, 33
  radial-flow 23, 26, 27, 33
  rectangular 16, 22, 34
  scaling-up 44, 46
Settling efficiency 23
Settling parameter 48
Settling process 13
Settling tank
  secondary 126
  vertical-flow 27, 28
Sherwood number 67, 94, 101
Simcar-type aerator 81
Similarity
  approximate 23, 39, 72, 94, 120, 169
  biological 143, 164, 165
  complete (Jull) 40
  dynamic 19, 39, 143
  economic 144, 175, 182
  geometric 8, 11, 21, 26, 27, 36, 39, 43, 63, 72, 79, 80, 81, 86, 87, 92–94, 96, 98, 100, 104, 106–108, 113, 124, 128, 135, 142, 148, 153, 154
  hydraulic 19, 23, 26, 39, 60, 61, 78, 159, 183
  kinematic 40, 143
  mechanical 63, 64
  partial 169, 186
  physico-chemical 143
  technological 144, 166, 187
  thermal 166, 183
Similarity theory 5, 31, 40

Simulation of trickling filter 74
Sludge age 165
Sludge blanket clarifier 56, 57
Sludge concentration, dimensionless 127
Sludge growth 127
Sludge loading 166
Sludge treatment 133
Sorption correlation 115
Sorption number 115
Sphericity 15
Stability of flow 29, 60
Stanton number 158–160
Stirring 53, 93, 113, 123, 140
Stokes law 14, 15, 48, 63
Strouhal number 45, 46, 151
Substrate concentration, dimensionless 127
Surface loading 38, 74, 75

Thickener 145
Three-dimensional hydraulic radius 80
Time ratio 23
Toxic effect 130
Trickling filter 74, 77, 126
Trivial modelling 37, 77, 164, 165, 167, 171
Tube separator 44
Turbulence 21, 31, 40, 42, 59, 61, 77, 89, 98, 101, 129, 151, 153

Velocity gradient 52, 53, 56, 57, 92
Venturi-type aerator 104, 106
Vertical-shaft aeration system 81, 119, 149, 151, 154, 157, 162, 163

Wall effect 170, 171
Wastewater treatment, advanced 69
Weber number 84, 93, 95, 97, 100, 102–104, 109, 121
Weissbach's coefficient 7
Wind effect 44

Yield constant 127

Zeta potential 49, 51